Office Ergonomics

Office Ergonomics

Karl H. E. Kroemer
and Anne D. Kroemer

London and New York

First published 2001 by Taylor & Francis Ltd.
11 New Fetter Lane, London EC4P 4EE

Simultaneously published in the USA and Canada
by Taylor & Francis Inc.
29 West 35th Street, New York, NY 10001

Taylor & Francis is an imprint of the Taylor & Francis Group

©2001 Karl H. E. Kroemer and Anne D. Kroemer

Typeset in Sabon by Deerpark Publishing Services Ltd.
Printed and bound in Great Britain by Biddles Ltd, Guildford and Kings's Lynn

British Library Cataloguing in Publication Data
A catalogue record for this book is available from the British Library

Library of Congress Cataloging in Publication Data
A catalogue record for this book has been requested

ISBN 0-7484-0952-1 (hbk)
ISBN 0-7484-0953-X (pbk)

For Hiltrud – on her 2001 anniversary

Contents

Foreword

Working at ease and efficiently

This is a "practical" book, but it is based on sound theory and research. It is written so that everybody can use it: you and I, the individual "office worker" or the boss, the equipment purchaser, the designer and architect.

We operate keyboards, check and make files, phone and fax, sit or stand, write and read, discuss and evaluate, prepare and make decisions. We all wish to be productive and successful, and most of us want to get along with everybody else in the office. Nobody wants to suffer from pains in the wrist or shoulder or back, and we all want to avoid "headaches", both literal and figurative ones.

We all have experienced those hours when were satisfied with the work, even liked to do it, and when it was actually fun — why shouldn't it be this way more often?

This book suggests how to set up the office, at home or at the company; how to select and arrange equipment and furniture; how to organize and pace the work; how to understand and interact with our colleagues at the office: how, in other words, to perform "at ease and efficiently". This is the motto of ergonomics.

People are different – how boring if we were all the same!

The recommendations in this book apply mostly to North American and European users. For example, the chair measures suggested here fit their bodies – but other populations with fundamentally different body sizes would need other ranges in furniture dimensions and adjustments. Furthermore, working postures and habits can differ strongly among the various peoples on earth. Preferences for colors, lighting, or temperature in the office vary from one continent and climate zone to the next.

How to use this book

You may use this book in three ways:

1. Read it straight through from beginning to end, as in a university course. The checklist at the end of each section should help you to consider whether you should change things in your office environment to "feel good".
2. Read a section of interest, think about applying that information, and proceed to the ergonomic recommendations that make use of that information. Do they make sense to you?
3. Use the index to look up a topic of concern to you.

Let us know

If you find that this book is of help, or that we should include something else you feel is important, write to Taylor & Francis, the publisher; or e-mail us directly at Kroemer@vt.edu. We would like to hear from you.

Karl and Anne Kroemer

1   Ergonomics applied to offices

Office design encompasses the design of whole workspaces, of individual workstations, and of their components such as computers or chairs. This has become a major task for the ergonomist, a person trained to design the work tools and to arrange the conditions in the office so that they

- enhance people's wellbeing,
- make work easy to do,
- allow people to perform efficiently.

Ergonomics defined

In the introduction to their 2001 book *Ergonomics – How to Design for Ease and Efficiency* the Kroemers gave the following definition: "Ergonomics is the application of scientific principles, methods, and data drawn from a variety of disciplines to the development of engineering systems in which people play a significant role. Among the basic disciplines are psychology, cognitive science, physiology, biomechanics, applied physical anthropometry, and industrial systems engineering".

Ergonomics (also called Human Factors or Human Engineering in the United States) is neutral: it takes no sides – neither employers' or employees'. It is not for or against progress. It is not a philosophy, but a scientific discipline and technology.

Ergonomics can make simple objects such as pencil sharpeners, wrist rests for the keyboarder, or telephones user friendly and easy to use. Of course, designing a public transport system, or a manufacturing plant, or the organizational set-up of a company is a much more complicated enterprise than, say, designing office lighting. Designing the organization of a large company sets up many people at different levels; all interact with each other, use various technologies, and have different tasks and responsibilities. Hence, designing such socio-technical systems requires the involvement of ergonomics at a high and complex level; this is often called 'macro-ergonomics', as defined by Hendrick and Kleiner (2001). It goes beyond the traditional 'micro-ergonomics' that concern specific features of hardware, software, job and environment. Of course, even designing a light pen or a computer mouse involves more than just considering hand size and mobility; so to label human engineering tasks either micro- or macro-ergonomics can be a matter of judgment.

Ergonomics applied to the whole organization directly or indirectly affects every employee, as discussed further in Chapter 2. Ergonomics impacts us every day: in the size and layout of our workspace (see Chapter 3), the heating and cooling of the office, the design and comfort of the chair and the keyboard. All of the human engineering discussed in Chapters 4–9 can make our daily work easier and more efficient.

Success of ergonomic effort is measured by improved productivity, efficiency, safety, and acceptance of the resultant system design and – last but truly not least – by improved quality of human life.

Humans run the office

The office is a work system that entirely depends on humans: without them, no work gets done. Therefore, ergonomics focuses on the human as the most important component of the office and adapts the office to the people involved. Such human-centered design requires knowledge of the characteristics of the people in the office, particularly of their dimensions, their capabilities, and their preferences.

People in the office have been in focus for at least 300 years, after Ramazzini stated in 1713 that workers who sat still, stooped, looking down at their work often became round-shouldered and suffered from numbness in their legs, lameness, and sciatica. Ramazzini generalized that "all sedentary workers suffer from lumbago," and he advised that people not sit still but instead move the body and "take physical exercise, at any rate on holidays". Ramazzini also took note of the writer's cramps from which scribes often suffered in the office (Wright, 1993, 180–185).

Today, musculoskeletal disorders in the hands and back, often together with eyestrain, are common complaints of people who operate computers in the office. Liberty Mutual Insurance Company reported in 1999 that hand, wrist and shoulder disorders were a fast growing source of disability in the American workplace, stemming in large part from the dramatic increase in the use of computer technology in the latter part of the twentieth century. According to Liberty Mutual (1999, p. 14), "as computers have become a staple in the workplace, work-related musculoskeletal irritation has increased." This development is disappointing, even deplorable, because it could have been averted by timely and proper application of ergonomic knowledge. Applying what we know today should help us avoid many of these problems in the future.

Ergonomics is not a new science, but rather relies on more than 100 years of physiological, psychological and engineering observation and research – even far longer if you consider Ramazzini's early efforts from the 1700s. A brief account of the development of ergonomics can be found in the Kroemers' 2001 book (previously cited) which reports that about a dozen other tomes on ergonomics were published since 1994 in North America alone. This count does not include books on more specific topics, such as biomechanics or repetitive strain disorders.

In today's offices, many work-related musculoskeletal disorders are computer-related injuries: for example, the widely known disease, carpal tunnel syndrome, is generally associated with keyboard use. However, until the 1980s most offices had no computers, and even today many offices exist without them. Office workers who do not use electronics (and those who do) often suffer from work-related discomfort, pain, and disease that since Ramazzini have been traced to

- lack of whole-body movement,
- unsuitable postures especially when maintained for long periods, often caused by ill-fitting furniture including chairs,
- physical overexertion of hand, arm, and shoulder by repetitive work.

Other problems are related to

- inadequate lighting,
- excessive noise, and
- stressful climate in the office.

Very important aspects of the working conditions include the power structure within the whole company or the office, relations between managers and employees as well as relations among employees, work autonomy and responsibility, job security, and other organizational aspects (organizational behavior is covered in some detail in Chapter 2). They can have overt negative effects on the office workers' attitudes leading to lack of motivation to perform, unwillingness to cooperate, dissatisfaction with work, unhappiness in general, even outright rage. There is good reason to expect that these psychological effects aggravate physiological problems like the ones listed earlier. Physiological and psychological traits are traditionally treated as separate academic topics but, in reality, they are inseparably intertwined and together determine a person's overall wellbeing.

Negative outcomes, such as the ones just mentioned, are at the top of what we perceive as pressing issues – and indeed they should be, because being injured at work, or being uncomfortable, or just not being happy with the job should not occur. On the positive side, knowing what is

- safe,
- easy, and
- agreeable

enables, allows, and obliges us to set up the office so that we are satisfied with the conditions under which work gets done and enjoy what we are doing, and therefore are motivated to perform well.

People do not come in one size, they do not have the same body proportions, and one chair size does not fit all. They do not all have the same preferences and dislikes. They are unique, they behave differently as Figure 1.1 depicts, and they want and deserve to be treated as individuals.

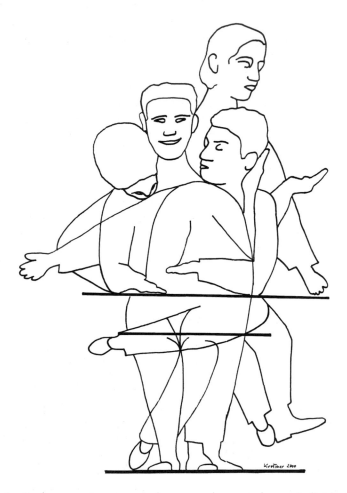

Figure 1.1   Each person is unique and wants to be treated as an individual.

Computers changed the office in the 1980s

Grandjean (1987) was apprehensive about the impact of the newly popular computer on the office worker. His particular concern was directed at the feared effects of radiation coming from the display, especially on pregnant women. Other worrisome matters included musculoskeletal overuse disorders due to repetitive activities, especially keyboarding, already experienced in the so-called repetitive stress injury (RSI) epidemic in Australia in the early 1980s. Another major topic was the luminance of electronic displays in combination with lighting of the office. The layout of the computerized workplace for suitable posture of the operator was still a point of discussion since early computers had been simply placed onto the existing desk or table originally designed for traditional office work.

While the fear of radiation damage to the computer operator subsequently diminished, all the other concerns were still very much alive in 1990 when Sauter and his co-authors addressed the issues of productivity and operator health in the computerized office. Office managers and designers had not foreseen that the computer would develop rapidly into a new work tool that was so distinctly different and would change the work in the office so profoundly that it deserved specific workplace layout as well as revised work procedures and new management attitudes. Who would have thought, around 1990, that the computer would so dramatically change the tasks and responsibilities of the office secretary: first into those of a word processing specialist, then of an administrative assistant? Who anticipated that every office staff member, the managers, even the boss (and all the professors) would do so much keyboarding themselves – work which, just a few years ago, would have been done exclusively by their secretaries? In fact, much of what was formerly office work is now done outside the old office; commonly on a laptop or handheld computer during commuting or travel, at times other than the old eight-to-five office hours, and increasingly in the home office.

Handheld and laptop versus desktop computers

Handheld palm-sized computers are growing ever more popular. In the beginning, their purpose was less for writing extensive documents than for brief notes and for communications with the outside. The handheld computer is replacing the notebook and calendar of yesteryear for many, and is taking over even many functions of full-fledged computers. In spite of their popularity and impressive features, many palm computers on the market as of this writing (in the year 2001) have design and use features that need improvements. Often the display is small and tough to decipher, buttons are difficult to operate, and the stylus is clumsy to use. Further developments will benefit from lots of ergonomic inputs; better human engineering will help propel the handheld computer as it forges inroads in offices and households.

The laptop computer is the direct high-tech descendant of the old typewriter. The laptop has a small keyboard with not many more keys than the typewriter. The display sits about where the platen of the type-writer used to be. The screen is placed appropriately for human vision, close to the keys (which are another important visual target) so that we look down at them all. (See Chapter 7 for more on this topic.)

The laptop's small size makes it possible to place it into many locations convenient at the moment. We can place it not only on our lap as we sit, but also on a crossed leg, set it on the arm rest of an easy chair, and of course position it on a table, desk, or any other feasible keyboard support. We can place it anywhere in a sit-down or on a stand-up work-station. We take it wherever we want to go. Thus, handheld and laptop computers allow us to change location and posture at a whim – they provide much more mobility and versatility than the bigger and heavier desktop computer. And most users become quickly accustomed to the smaller size and peculiar resistance of the keys. A larger display and an additional keypad as well as other peripherals can be connected to the portable computer when it is set up in the stationary office.

For travelling and easy use outside the office, the handheld and the laptop computer are feasible tools; indeed, there is nothing wrong with using a well designed portable computer in the conventional office – in fact, it facilitates moving about rather than sitting still. Motion is desir-able (see Chapter 4). With a bit of imagination, we can even see ourselves in a future work environment in which we are free to move about, sit or stand as preferred, because our input to the system will be by voice and gesture or, if manual, by wireless keyboard or other gadget – and the display may be flat, head-mounted or even virtual, floating in space for easy placement wherever we want it.

The home office

The term "home office" still has the connotation of a corner in the den, a spot in the kitchen or, at best, a spare room, used to do occasional paperwork for an hour or so, as the need arises. That inferred meaning also implies the use of furniture that is not specifically designed for office work, like a folding chair at a card table or an easy chair in that corner of the den. However, as the occasional work changes into regular schedules of workdays with long hours of effort, the room housing the home office must be carefully selected, suitably cooled or heated, properly insulated and protected from noise and interference, and well lit. And all the furniture must be carefully selected as well.

Let us not fall for unsuitable shelf-type set-ups, as often advertised for the home office: many are badly designed and not appropriate for common use, as opined by the human factors expert. Instead, we should carefully select a workstation that we can adjust to fit our own body size and our own preferred working habits; ideally, suitable for alternating sitting and standing. If your local stores do not offer appropriate furniture, check catalogues and the Internet. Appealing, well-designed adjustable workstations are widely offered, at very reasonable prices. Consider using, instead of a desktop computer, a laptop. Its small keyboard requires much less hand and arm movement. This might help to avoid repetitive trauma disorders (see Chapter 6). And you can place the laptop wherever you wish, use it sitting or standing up, and take it with you when you travel.

If you decide to use a multi-piece desktop computer set-up, make sure that your furniture allows you to put the display directly behind the keyboard, not far above, not to the side. The box with the central processing unit (CPU), the innards of the system, and the printer can be placed a short distance away – you will not need to use them that often, and it is actually a good idea to get some body exercise by reaching over, even getting up to change the disk or retrieve the printout.

Make sure that you have ample room to stretch your legs. And, of course, you should have a chair that really fits you and that integrates well with the rest of the office furniture. The home office deserves full ergonomic attention because it can and should be personalized to suit you, in the interest of both your wellbeing and your performance. (More about the home office in Chapters 2, 3, and 4.)

Childrens' computer workstations

The computer has not only changed the working conditions for many adults, but is now a major toy for play and a learning tool for children at home, and an essential instructional instrument in the schools, both elementary and secondary. The ergonomist notices with great concern that small children are made to sit at workstations that were originally designed for adults, with seats and tables that do not fit their bodies, mostly because the furniture is too big (Hedge et al., 2000; Saito et al., 2000; Straker et al., 2000). The situation is often exacerbated by the old mistaken practice of putting the monitor on top of the CPU (central processing unit) box. So the children must crane their necks to look up to the display even while seated on a chair that is too tall for them – their feet usually do not reach down to the floor. Furthermore, the regular computer keyboard, with its large size and excessive number of keys, does not fit the small child's hands. Instead of using a desktop computer, consider whether a laptop might not better suit the child, both with regard to the size of the keyboard and the location of the screen. Not often available yet in furniture stores, but offered in catalogues and over the Internet, are computer workstations that fit the body size of children and can be adjusted to their quickly changing dimensions and habits.

Tasks in the office

There are many offices that differ in size, location, layout, organization, and purpose, and there are many different ways to categorize the work that is being done in them. Since this text is concerned with the ergonomic aspects of task performance and resulting workload, we divide office tasks into four gross divisions of major work categories:

1. Prepare texts and correspondence, such as letters and invoices, by typewriter or computer.
2. Keep track of and record payments, schedules, and events.
3. File and retrieve material by hand or electronically.
4. Talk with others by phone or face to face.

Of these activities, only talking with others has little dependence on the use or non-use of computers. Filing and retrieving materials by hand is, in the interest of getting up and moving about instead of sitting still, a desirable activity – especially if the filing cabinets are located away from the workstation. (For the industrial engineer it is curious to realize that it is advantageous not to have all work tools within close reach – closeness used to be a goal of efficient workplace design – because many office people need more whole-body physical exercise.)

The first two tasks, preparing texts and correspondence and keeping records, are indeed quite different in the computerized and the non-electronic office. The physical effort per keystroke is much larger on a mechanical typewriter than on an electronic keyboard because of the larger key travel and the larger force needed to overcome the dynamic resistance; the resultant work is several times that needed to trigger the key on a computer. This large effort, multiplied by the number of keystrokes completed, was the reason why so many operators of mechanical typewriters suffered from musculoskeletal overexertion injuries, reported since the 1920s. (More on this topic in Chapters 5 and 6.)

However, musculosketelal overexertion injuries also occur to operators of electronic keyboards; common experience shows that, per unit of working time, the number of key activations is even larger on computer keyboards than with mechanical typewriters. Apparently, the number of digit motions exceeds the capabilities of many computer operators' hand tendons – and, as mentioned earlier, many of us other than the employees formerly designated as secretaries currently do keyboarding on a daily or routine basis.

Working hours and power naps

Concurrent with the spreading use of the computer is the reality of extended working hours per day. Our daily work schedules have expanded in recent years, at least in the USA; the proverbial nine-to-five workday in the office, 5 days a week, is no longer the limit for many people. More working hours per day, in addition to work on weekends, place a heavy physical and mental load on a person and shortens the time available for recuperation, relaxation, and doing what we like to do.

There is no hard limit for the working time that applies to every task and every person. The literature is in agreement about the hypothesis that 8 hours during daylight is a suitable work duration for most office tasks – see, for instance, the overview in Kroemer et al. (2001) or, more specifically, Tepas (1999) and Tepas et al. (1997) for chapters on work schedules and performance. We know from our own experience that such 8-hour shifts can be almost insufferably long if the workload is high and continuous, or that we can perform well for a longer work shift when the load is light and intermittent and interesting. Most of us are especially productive or motivated when we can set our own time schedule. Some people, like journalists or poets, habitually work during unusual times of the day, and for long stretches of time. Yet, for us regular office workers, 8 hours are enough, and many of us do best during the morning while some people are highly productive during the afternoon.

For most, a short break, possibly even a brief nap, after lunch is important. The literature acknowledges the existence of the after-lunch dip but cannot pinpoint one certain definitive chrono-biological reason for it (an example of a missing scientific explanation for a common phenomenon). For some of us, the need for the post-lunch nap is related to the desire for peaceful digestion of food and drink, together – quite reasonably - with the habit of recharging one's batteries for more work. Many of us are forced to conceal taking our forty winks although napping openly (as reportedly done, for instance, in Japan in many companies) would be more beneficial. Interestingly, some companies appear to be acknowledging the benefits of a nap, making nap-rooms available to employees. A large advertising agency in Chicago, for example, installed nap rooms in its flagship office building in the late 1980s, letting employees sign up for the rooms in half-hour increments. (Of course, although some companies do sanction nap times, getting your boss to approve of the daily snooze is a different matter). Winston Churchill, Napoleon Bonaparte, Albert Einstein, and some US Presidents (Kennedy, Reagan, and Clinton) habitually took naps, reports Jane Brody (New York Syndicate, in the January 11, 2000, ROA Times, page E 3).

There are convincing physiological and psychological reasons for taking short breaks often, in addition to a longer pause about halfway during the work shift. Older and recent research has shown, again and again, that frequent short interruptions of physical work restore the ability to continue working at a high level. (Sustained keyboarding is a hard effort for the musculoskeletal components of the hand and forearm and necessitates occasional breaks for maintained health.) Taking a pause of 5 minutes every half hour is better than 10 minutes once an hour even though the total break time is the same (Balci et al., 1998; Kroemer et al., 1997, 2001; Neuffer et al., 1997).

Good office ergonomics makes good economics

As already mentioned, Liberty Mutual Insurance Company reported in 1999 on a test of the theory that, by giving employees more control over their environment and a better understanding of ergonomic principles, their performance would improve and their health problems diminish. The results of the 18-month study confirmed the expected: combined with ergonomic training, the flexible workspace increased individual performance and group collaboration. This was accompanied by a nearly one-third reduction in back pain and a two-thirds reduction in upper limb pain among the employees who had more control over their environment.

This study confirms similar findings made a decade earlier. Planning a good work environment that facilitates employee and organizational effectiveness is a good investment, as Francis and Dressel reported (Sauter et al., 1990, 3–16). After comfortable, adjustable, task-oriented ergonomic workstations were installed, the performance and the satisfaction of office employees were significantly improved. The payback for the expenses incurred to improve the conditions took less than 11 months. In the same book (Sauter et al., 1990, 49–67) Dainoff reported on a carefully controlled laboratory study with two computer workstations, one poorly laid out, one optimal in terms of ergonomic design aspects. After a 5-day test period, the test participants unanimously preferred the better workstations, felt more comfortable and performed better. Part of this outcome may be attributed to the so-called Hawthorne effect: it is related to the simple (and correct) perception that the one set of office equipment was obviously better than the other one – but is that not real life?

One design does not fit everybody

This text contains specific design recommendations, for example, in chair dimensions, which have been derived to fit mostly North American and European users. Their body sizes are fairly well known, but those of most other populations on earth are not – see the compilations of international anthropometry by Pheasant (1996) and Kroemer et al. (1997, 2001). People with basically different body sizes would need their own specific ranges in furniture dimensions and adjustments. Furthermore, working postures and habits can differ significantly among the various people that inhabit the globe, as described, for example, by Kroemer et al. (2001) and Nag et al. (1986). Also, national regulations and standards can be quite different, as a comparison between the fairly similar European (Cakir, 1999; 2000a,b; Cakir and Dzida, 1997; Stewart, 2000) and the US standards (e.g., ANSI, ASHRAE, OSHA) easily demonstrates.

Humans live in diverse social and organizational structures, have their special ways, preferences and dislikes, and are accustomed to their own climates, sounds and colors. Therefore, the recommendations given here most likely need to be modified to suit users in parts of the globe other than North America and Europe.

Office design combines science, technology, and art

Kleeman and his co-authors (1991) provided an overview of the development of office design during the last approximately 150 years. Intuition and artistic ideas were the main drivers for the various design principles of the office as a whole: from the small households of several clerks in their office suites to the vast open spaces of the "paperwork factories". A next step was the landscaping of large offices, indicating a new notion of relaxed human relations; which, however, was soon replaced by the tough concept of modular spaces and cubicles, with more luxurious private offices for those lofty souls ranking higher in the management hierarchy.

These approaches to office designs are not only intuitive, however, but also reflect philosophical and social ideas. They are also strongly based on the current and available technologies, such as heating and cooling of the rooms, and on the predominant work methods: from writing by hand to typewriting to computer use, from telegraph to pneumatic tubes to telephone to wireless electronics. Technology is always in flux – what is current today may be thoroughly outdated tomorrow. New technologies can be forgotten or become common almost overnight, as the telegram and email demonstrate.

Technology and science also have influential roles in the design of equipment in the office. Reading books by Grandjean (1987), Sauter et al. (1990), Kleeman et al. (1991), Raymond and Cunliffe (2000), Tilley (1993), and Lueder and Noro (1994) makes it clear that science-based biomechanical knowledge is not sufficient to design a comfortable office seat; lighting engineering alone cannot make an office appealing to the eye; air-conditioning per se does not make all office workers satisfied. Ergonomics (also called human factors engineering in the USA) incorporates the scientific and engineering disciplines that are concerned with the human at work, but the ergonomist still must work with the artistic, intuitive, daring, and speculative designer of furniture and office spaces to create that new office.

Science influences office design by providing theories about efficient and humanistic work systems and models. Science also provides methods and techniques to measure the outcomes: *subjectively* such as by satisfaction or motivation, and *objectively* by such metrics as productivity or cost effectiveness.

The literature also shows that, in the past, it was management that determined, primarily according to economic considerations, which one of the many possible office designs was to be selected. This probably was, and is, a reasonable process – but the assessment of what is good for the human working in that office is (or should be) an important component of the decision making. Still, there are many irrational, intuitive, wishful, and unsupported ideas that have influenced the judgment, such as "we think faster on our feet than on our seat", "posture affects mood and willingness to work", "wall colors with red in them make people excited and aggressive". Shall we then return to standing upright in the office, and paint the walls gray-green?

No doubt there are visionary and speculative traits in office layout, and without them some daring new designs and insights would not exist. Yet, people are the doers of office work, they derive new ideas and procedures, they determine success or failure; they are the drivers and pilots, and – as with the cockpit in cars and planes – their office must be designed around them. Ergonomics supplies the information needed for that human-centered design.

References

Balci, R., Aghazadeh, F. and Waly, S. M. (1998). Work-Rest Schedules for Data Entry Operators, in S. Kumar (Ed.), *Advances in Occupational Ergonomics and Safety*. Amsterdam: IOS Press, 155–158.

Cakir, A. E. (1999). Human-Computer Interface Requirements, in W. Karwowski and W. S. Marras (Eds.), *The Occupational Ergonomics Handbook*. Boca Raton, FL: CRC Press, 1793–1812.

Cakir, A. E. (2000a). International HCI-Standards on Keyboards and Other Input Devices, in *Proceedings of the XIVth Triennual Congress of the International Ergonomics Association and 44th Annual Meeting of the Human Factors and Ergonomics Society*, Santa Monica, CA: Human Factors and Ergonomics Society, 6406–6409.

Cakir, A. E. (2000b). International HCI-Standards for the Workplace and the Work Environment, in *Proceedings of the XIVth Triennual Congress of the International Ergonomics Association and 44th Annual Meeting of the Human Factors and Ergonomics Society*, Santa Monica, CA: Human Factors and Ergonomics Society, 6410–6414.

Cakir, A. E. and Dzida, W. (1997). International Ergonomic HCI Standards, in M. G. Helander, T. K. Landauer and P. V. Prabhu (Eds.), *Handbook of Human-Computer Interaction* (2nd ed.). Amsterdam: Elsevier, 407–420.

Grandjean, E. (1987). *Ergonomics in Computerized Offices*. London: Taylor & Francis.

Hedge, A., Barrero, M. and Maxwell, L. (2000). Ergonomic Issues for Classroom Computing, in *Proceedings of the XIVth Triennual Congress of the International Ergonomics Association and 44th Annual Meeting of the Human Factors and Ergonomics Society*. Santa Monica, CA: Human Factors and Ergonomics Society, 6296–6299.

Hendrick, H. W. and Kleiner, B. M. (2000). Macroergonomics: An Introduction to Work System Analysis and Design. Santa Monica, CA: Human Factors and Ergonomics Society.

Kleeman, W. B., Duffy, F., Williams, K. P. and Williams, M. K. (1991). *Interior Design of the Electronic Office*. New York: Van Nostrand Reinhold.

Kroemer, K. H. E., Kroemer, H. J. and Kroemer-Elbert, K. E. (1997). *Engineering Physiology* (3rd ed.). New York: Van Nostrand Reinhold–Wiley.

Kroemer, K. H. E., Kroemer, H. B. and Kroemer-Elbert, K. E. (2001). *Ergonomics: How to Design for Ease and Efficiency* (2nd ed.). Upper Saddle River, NJ: Prentice Hall.

Liberty Mutual Research Center for Safety and Health (1999). *From Research to Reality. Annual Report*. Hopkinton, MA: Liberty Mutual Research Center for Safety and Health, p. 14.

Lueder, R. and Noro, K (Eds.) (1994). *Hard Facts About Soft Machines. The Ergonomics of Seating*. London: Taylor & Francis.

Nag, P. K., Chintharia, S. and Nag, A. (1986). EMG Analysis of Sitting Work Postures in Women. *Applied Ergonomics 17*, 195–197.

Neuffer, M. B., Schulze, L. J. H. and Chen, J. (1997). Body Part Discomfort Reported by Legal Secretaries and Word Processors Before and After Implementation of Mandatory Typing Breaks, in *Proceedings of the Human Factors and Ergonomics Society 41st Annual Meeting*. Santa Monica, CA: Human Factors and Ergonomics Society, 624–628.

Pheasant, S. (1996). *Bodyspace* (2nd ed.). London: Taylor & Francis.

Raymond, S. and Cunliffe, R. (2000). *Tomorrow's Office* (2nd ed.). London: Taylor & Francis.

Saito, S., Sotoyama, M., Jonai, H., Akutsu, M., Yatani, M. and Marumoto, T. (2000). Research Activities on the Ergonomics of Computers in Schools in Japan, in *Proceedings of the XIVth Triennial Congress of the International Ergonomics Association and 44th Annual Meeting of the Human Factors and Ergonomics Society*. Santa Monica, CA: Human Factors and Ergonomics Society, p. 7658.

Sauter, S., Dainoff, M. and Smith, M. (Eds.) (1990). *Promoting Health and Productivity in the Computerized Office: Models of Successful Interventions*. London: Taylor & Francis.

Stewart, T. (2000). Ergonomics User Interface Standards: Are They More Trouble Than They are Worth? *Ergonomics 43*, 1030–1044.

Straker, L., Harris, C. and Zandvlioet, D. (2000). Scarring a Generation of School Children Through Poor Introduction to Technology in Schools, in *Proceedings of the XIVth Triennual Congress of the International Ergonomics Association and 44th Annual Meeting of the Human Factors and Ergonomics Society*. Santa Monica, CA: Human Factors and Ergonomics Society, 6300–6303.

Tepas, D. I. (1999). Work Shift Usability Testing, in W. Karwowski and W. S. Marras (Eds.), *The Occupational Ergonomics Handbook*. Boca Raton, FL: CRC Press, 1741–1758.

Tepas, D. I., Paley, M. J. and Popkin, S. M. (1997) Work Schedules and Sustained Performance, in G. Salvendy (Ed.), *Handbook of Human Factors and Ergonomics* (2nd ed.). New York: Wiley, 1021–1058.

Tilley, A. R. (1993). *The Measure of Man and Woman. Human Factors in Design*. New York: The Whitney Library of Design.

Wright, W. C. (1993). *Diseases of Workers. Translation of Bernadino Ramazzini's 1713 De Morbis Articum*. Thunder Bay, Ontario: OH&S Press.

2 Relations among people in the office

Overview

This chapter deals with the primary macro-ergonomic issues (Hendrick and Kleiner, 2000; Kleiner and Drury, 1999) that largely determine the role of every person in the system:

- How the individual employees feel about their company – "this is a good place to work."
- How much they want to be involved – "I don't know anything, I just work here."
- How important they consider their work – "if I don't show up, this office will come to a standstill."
- How people get along with each other at work – "my buddies at work, almost like family."
- How much effort they put into their work – "good enough for government work."

Organizational behavior

The field of study that helps us understand and deal with the interpersonal and organizational challenges in the workplace is called *Organizational Behavior*. While knowledge of many of the quantitative micro-ergonomic aspects we cover in Chapters 4–9 is essential, we also need to have the sensitivity to apply these skills in the real world of our organization. Before we delve into the inner workings of an organization and the psyches of the individuals employed there, let us briefly make our case for why we think it is valuable to do so.

Now that the lean and mean years of the last century are behind us, more and more companies are trying to keep their employees happy – satisfied and motivated. This is not sheer altruism at work; instead, organizations are simply recognizing that improving employee satisfaction will improve bottom line profit. Happy employees, they find, are productive; they treat customers better, work harder, and even take fewer sick days. Moreover, they tend to stay with the organization, which reduces one of the most significant costs – employee turnover related to unhappy employees. Not only does employee exodus carry a huge expense in terms of recruiting and training new employees, but a new hire accomplishes substantially less initially – in some studies, more than one-third less in the first 3 months – than an experienced worker. The outright expense of replacing a valuable employee can range from half to several times a year's pay. Finally, there is the less tangible cost of unhappy employees who stay on the job; not only should we consider their low productivity and poor customer service, but we must also include the stress (and, potentially, serious stress-related illnesses) that they suffer as they trudge through a dreary work routine.

The elements of an organization

To understand an individual in her or his job, we need to understand not just the person, but the environment – the organization – in which he or she operates. And to understand an organization, we must consider all of its components. Organizations are networks of related parts, and all of the elements work together to enable the organization to function as a whole. This chapter covers macro- and micro-views of the organization, starting with the elements that define organizations and ending with the people that work there.

The basic organizational model, shown in Figure 2.1, depicts the internal and external elements that define an organization. Six structural elements interact to define a company: strategy (the company's plan for success), structure (corporate hierarchies), policies and procedures (company rules and regulations), systems (methods of allocating,

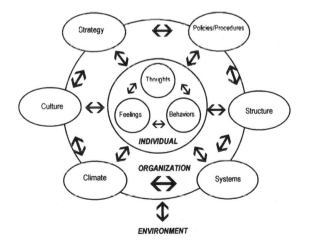

Figure 2.1   Basic organizational model.

controlling, and tracking corporate resources), climate (employees' feelings about the company), and culture (behaviors and feelings within the company). Each is explained further in this chapter.

The human is in the center

As shown in Figure 2.1, the individual is at the center of the organization's operations, because ultimately, companies are made not of theories, structures, and machines, but instead of living, breathing – and interacting – people. People affect the organization and the way it functions and, in turn, the organization affects these individuals. Why people act the way they do – and what we can do to keep them satisfied and successfully interacting with each other at work – is a fascinating puzzle that has kept psychologists and behaviorists busy for many decades.

Behavior

Many theories exist to explain an individual's behavior, including the APCFB Model (the acronym stands for assumptions, perceptions, conclusions, feelings, and behaviors). This model posits that a person's closely held assumptions color the perceptions of a given event, leading to highly individual conclusions; these in turn cause feelings which result in behaviors. Since everyone has a different set of assumptions, individual behaviors or reactions even to one common event can be widely divergent.

Motivation

The attempt to understand the motivation of employees underlies behavior theories: motivation is the attitude toward attaining a goal. The most widely cited original theories to explain behavior and motivation include Maslow's hierarchy of needs and Alderfer's ERG theory. The basic premise of both theories is that individuals continuously strive to satisfy certain needs, and that this quest in turn drives our behavior. Another well-known explanation is not internal or need-based but rather external; called the reinforcement theory, at its core is operant conditioning, originally conceived by B.F. Skinner. Reinforcement theory's three components are stimulus, response, and reward; the stimulus invokes a response, which is subsequently rewarded (or punished).

Job satisfaction

Job satisfaction is closely correlated to motivation and, naturally, several theories exist to explain the degree of pleasure an employee derives from his or her job. Some approaches postulate that job satisfaction is determined within the individual, when a person's physical and physiological needs are met; others center on the external factors of social comparison. Work conditions also may influence job satisfaction; Herzberg's two-factor theory is the most famous of the work condition theories (see Figure 2.3) (Herzberg, 1966). Herzberg believed that certain job content factors (like recognition and achievement) were positively associated with satisfaction. He postulated that other job conditions (like pay and the office environment) could generate dissatisfaction if perceived as negative.

Ultimately, all of the above theories contribute to some extent to the explanation and understanding of motivation and satisfaction. All are covered in some detail in subsequent pages of this chapter.

Stress

Stress is a universal phenomenon; it can occur at all organizational levels and in all jobs. Stress refers to an individual's psychological and physiological responses to environmental demands, and it carries at times severe consequences in terms of both physical and mental health. Reducing stress is critical for the wellbeing of employees and accordingly for the continued operational soundness of an organization. Techniques for coping with stress include setting and rearranging priorities, releasing emotions through actions or discussions, taking time off, moving on to new activities, or learning to accept a given situation. Some companies have implemented formal stress-reduction programs to help combat this invasive problem.

Communications

Communication is key in any relationship between persons; the relationships we forge at work are no exception. It is through communication that we establish hierarchies, foster motivation, exchange information, and express our feelings. One formal method of communication singled out for further examination in this chapter is the performance appraisal. In theory, an appraisal serves all the functions of communication; in practice, performance appraisals tend to be feared and avoided by both managers and employees. Yet they can be extremely valuable and in fact necessary when done properly. To make them most effective, appraisals should be done regularly and according to a set schedule, have tangible results directly tied to the outcome of the appraisal, and be conducted interactively, with both manager and employee participating in the discussions. Additionally, many companies utilize peer appraisals and upward (subordinate-to-boss) appraisals to glean colleagues' impressions of the individual and fully gauge his or her contribution to the organization.

Quality of life – at work and off work

Striking a balance between work and life outside of the job is a final consideration of this chapter. A leanly staffed company in a tight labor market sets up a situation in which the employees might well be overloaded with work. Technological developments that connect us to work around the clock (such as pagers and cell phones and handheld computers) also influence the burden many employees feel. Organizations again realize that overtaxing employees may lead to reduced productivity and even employee exodus; as a result, a number of companies are taking measures to keep employees from feeling overwhelmed by work. These measures include lengthy and uninterrupted vacations, limited meetings, flexible schedules, and employee sabbaticals.

> You may skip the following part and go directly to the Ergonomic Recommendations at the end of this chapter – or you can get detailed background information by reading the following text.

What, exactly, is the organization?

As depicted earlier in Figure 2.1, the six main components of an organization are strategy, structures, policies and procedures, systems, climate and culture. All interact with and affect one another, and as the environment changes, they too must adapt.

- *Strategy* refers to the plan – stated or implicit – that the company has for success against its competitors. It guides the company's operations and determines specific tactics that the company will use to meet its overall strategy. Examples include the advertising agency that offers complete integrated marketing campaigns rather than just the advertisements themselves; the plumbing service that pledges to dispatch a plumber within 2 hours of a service call; the grocery store chain that offers a wider selection of fresh produce and gourmet items than competing stores.
- *Structures* outline the hierarchies within an organization. In larger companies, they are usually depicted in organizational charts, which are detailed structural diagrams – most of us have seen several versions at some point in our careers. An organization's structure determines accountability and authority within its ranks; essentially, it defines the official relationships that exist between employees. Each level in a structured organization has its own degree of authority and responsibility; as you move up in the hierarchy, you increase your level of authority and responsibility. In general, each employee should

only be accountable to one boss; this is known as the unity of command principle.

Structures traditionally fall into seven main categories; these include *function, product, customer, geography, division, matrix* and one called *amorphous*, if none of the above applies or works. *Function* means that work is divided by specific task, like finance, accounting, marketing, etc., and the various departments carved out by the structure generally report to an executive vice-president. *Product* means that the staff is divided by all functions necessary to produce and sell a given product or brand; this structure usually features product or brand managers who are entrepreneurs running their division almost like a separate company. Many consumer packaged-goods companies are organized in this manner. The *customer* structure divides all functions by customer need; this is the type of structure often found in service industries. *Geographic* structures organize work by location, and often, regional offices are set up to manage the business in a specific geographic area. A *division* structure means that each business unit operates like an independent organization under the overall umbrella of a parent corporation. Although a division may run itself almost autonomously, financing is often handled by the parent company. The *matrix* structure is unique in that an employee may have more than one boss because two or more lines of authority exist – this then is the only structure that violates the unity of command mentioned above. The matrix structure is generally reserved for organizations that feature complex and time-consuming projects that call for specialized skills. A computer manufacturer or consulting firm might have a matrix structure. *Amorphous* means that the organization has no formal structure; instead, employees forge and dismantle reporting relationships as needed. In reality, most larger companies use a mix of operational structures; accordingly, their structure is hybrid.

Employees within any of these structures are generally distinguished as *line* or *staff*. Line employees are those who are directly involved in producing the organization's products or services or meeting its primary goals, while persons who support these functions (through administration, advice, or general support) are referred to as staff employees.

- *Policies and procedures* are the rules and guidelines that govern a firm's conduct. Policies are official rules and, in larger companies, are often formally written up in detail in an employee handbook. Examples of policies include the amount of medical benefits a company provides, retirement plans available, or the number of paid vacation days an employee receives. Procedures are often not documented, but are nevertheless widely understood by employees and are generally applicable to routine tasks. An example of a procedure might be how often e-mails are checked or whether or not coffee is provided free of charge in the employee lounge.

- *Systems* are developed by companies so that money, people, and things (machines, equipment, supplies) are properly allocated, controlled, and tracked. Systems fall into several categories of distribution and management, including money (accounting, investment, and budgeting), object (inventory and production), people (human resources, employee appraisals), and future (strategic planning, business development, marketing planning).
- *Climate and culture* are closely related. Climate refers to the emotional state of the people in an organization; how they individually feel about the company, their coworkers, and their jobs. (Note that engineers use the term climate in its physical sense, referring to temperature and humidity and the like – see Chapter 9.) Culture is more of a group phenomenon; it relates to the behaviors, beliefs, values, customs, and ideas that an organization encompasses. (Of course, use of the term culture here has little to do with its historical understanding.)

Consider two advertising agencies; one a large and highly traditional company headquartered in a metropolis, the other a small boutique agency in a smaller town. Here are some of the characteristics of each: the first agency is 60 years old, has many long-standing blue-chip clients, and is highly structured, with formal dress codes, office procedure manuals, and written policies for all conceivable situations. Internal memos follow a set format and are heavily scrutinized at several administrative levels before they are issued; annual retreats and parties are scheduled and planned many months in advance. The second agency is 10 years old, with only a handful of young partners who employ highly creative individuals in an entirely unstructured office. Shorts and baseball caps are standard attire, no handbooks exist, and decisions are made quickly and with virtually no bureaucracy.

Clearly, each of these companies – although both operate in the same industry – have very different cultures, with widely diverse values and behaviors, and hence with different climates. Employees in a company acquire its culture when they first begin working there; they become socialized into the broader group by interacting with the different members. Culture can be communicated subtly to new employees or loudly, with no ambiguity.

Climate and culture have become more important than ever, with management increasingly realizing that these elements are major determinants of employees' overall happiness within an organization. (That was well known before, but then apparently forgotten in the lean and mean decade of the late 20th century.) In this day of casual work environments, telecommuting, job-sharing, and the emerging emphasis on individuality, companies' cultures vary more widely than ever before – and are more important to people than ever before.

Climate and culture have become valuable bargaining chips to attract and retain employees. Recall the example of the advertising agencies outlined above. Each company attracts a different type of employee, and prospective recruits should understand the organization's culture before signing on. Especially in today's job market, where labor is tight, employees can choose company cultures where jeans are appropriate office attire, pets can accompany their owners to work, on-site massages are routinely provided, and pool tables and nap rooms are available for employee time-outs. Both the potential employee and the potential employer are well advised to carefully scrutinize the perceived fit of an individual within an organization: if a person does not appear to match the corporate culture, and vice versa, then an employee/employer relationship is often not effective and not advisable.

The individual

This brings us to the inner portion and most important feature of our basic organizational model: the individual. Today when a company's stock market value is determined, hard assets like property, plant and equipment generally make up one third to one half of its value. The remainder is made up of soft attributes that traditionally received far less attention from shareholders: patents, customer base, and employee satisfaction. As a result, there is more incentive than ever to examine the inner circle of the diagram in Figure 2.1 to assess employees' state-of-mind and find out what influences it. Human resources are a company's most valuable assets.

Individuals are unique. All of us are different; our upbringing, environment, experiences, and personalities all make us special. Getting along with people is fraught with difficulties even in the best of circumstances, so throwing groups of people together in a work environment and expecting everyone to get along is unrealistic at best. Add to that the tensions inherent in a job – stressful or monotonous duties, long hours, and imposed reporting relationships with people you may or may not like – and problems are virtually inevitable. The popular media reflects this reality; there are shelves filled with books that provide instructions on dealing with employees and bosses, magazines publish countless articles on dealing with difficult colleagues, even movies cover the treacherous terrain of interpersonal relationships on the job.

To gain insight into why people act the way they do on the job, we now take a look at some of the better known theories of behavior, motivation, and job satisfaction. Please note that this review is by no means comprehensive; instead, it provides a glimpse of some of the most well-known alternative views.

What, exactly, is behavior?

The APCFP Model

Many textbooks (e.g., Greenberg and Baron, 2000; Landy, 1989; Muchinsky, 2000) refer us to some form of the APCFP Model; this model offers an explanation of how an external event can cause an employee's behavior. The letters in the acronym stand for assumptions, perceptions, conclusions, feelings, and behaviors.

The model essentially postulates that the assumptions a person holds are an intricate part of the person's overall makeup; these assumptions are closely held beliefs that we have about the way the world and the people in it should or ought to be. To a large degree, many of the assumptions we hold dear were created or at least strongly influenced early in our lives by parents and peers. These assumptions make up our value system.

When an external event occurs, we see it through our own filters that influence how we view or perceive the event. Our filters include internal defense mechanisms that act to protect us from psychological damage. These filters often prevent us from having an accurate reading of external events – and of other people. When an event occurs, our existing value system and the filters we have subconsciously created shape our view of the event into a given perception, which may differ significantly from reality. Yet, our perception leads us to draw conclusions, which in turn lead to feelings, and these feelings then cause our behavior. Behavior includes doing or saying something, and this behavior may at times seem wildly out of context or proportion to the actual event, depending on the assumptions (and filters) that we held to begin with.

Looking more closely at these assumptions, we can further classify them based on how deeply rooted they are. *Expectations* are most easily changed; *beliefs* are more deeply rooted, and *values* are closely held assumptions; values are so strongly felt that they may not be changed at all, or only slightly modified over long periods of time.

Understanding and then tapping into the value systems of employees can be extremely effective. Doing so is often called empowerment. Let us say that a given employee treasures her creativity but is given little leeway to express this creativity at work. If her manager were to revise the job to allow her to express this creativity, the employee will most likely become far more productive and happy at work. This is called goal congruence – the employee's goals and her manager's goals, in this instance, become equivalent or congruent. Goal congruence among the employees of a department or an organization make the group far more productive.

To illustrate this model, let us take an example. An insecure worker with a limited marketing background is promoted to marketing manager and asked to supervise the creation of the company web site. When a web designer from an outside agency arrives to present the new web site, considered by others to be brilliant, the newly promoted manager feels threatened by the web master's knowledge, while realizing that she herself ought to be more skilled in web design. To her staff's surprise, she rejects the proposed web site out of hand in spite of its merits. Finding fault with the proposed web site allows her to avoid confronting her own perceived ineptitude in web design (her filters helped protect her self-esteem). In the meantime, the new manager's staff is left with severe doubts about its on-line presence.

To augment our example, if company management had assured the new manager that her lack of on-line knowledge was not a problem and that she would be provided with appropriate training, she may have relaxed her expectation that she should excel in the field. Instead, her defense mechanisms caused her to act in a way that appeared irrational or out of proportion to her staff and her web design agency. Her resulting behavior caused an unfortunate consequence that may cost the company significant money in lost sales since what might have been a solid on-line site was halted.

As another illustration, consider a financial analyst who works at a mid-sized company that boasts a well-developed intra-office computer system. This analyst was once a happy, energetic, and productive employee whose mood and energy of late have taken a dramatic downturn. His wife has just had their first baby, and he has a very strong system of family values. While she plans to take a 3-month maternity leave, he would like to spend more time at home with his expanded family. He has a very long commute to work and generally spends ten or more hours a day working and driving to work. His manager sees the toll leaving his baby and wife on a daily basis takes on his mood and productivity. Their goals are congruent; both want to find a way to ease the analyst's separation pains and both wish to restore his former cheerfulness and productivity. Together, the analyst and his manager work out a schedule whereby he can work out of his home 2 days a week by telecommuting and using his home computer to tap into the computer systems at work. Moreover, on the three remaining weekdays, he can adopt a flextime schedule whereby he arrives at work at 07:00 and leaves by 15:00. In this way, he can avoid much of the traffic that snarls and complicates his daily drive and spend more time with the new baby. Within just weeks of implementing these changes, the financial analyst's work and attitude have greatly improved.

Summarizing what we know about behavior

We all have closely held beliefs about how the world functions, how the people in it function (or how they should function) and how we fit into this system. Beliefs create filters through which we see and perceive everything around us. Our filters include internal defense mechanisms that protect us from psychological damage. But these filters may prevent us from having a correct understanding of external events and of other people. Our perception leads us to draw conclusions, which in turn lead to feelings, and these feelings then cause our behavior. This behavior may at times seem wildly out of context or proportion to the actual event, depending on the assumptions and filters that we hold.

Our *expectations* are most easily changed because they are least entrenched. *Beliefs* are more deeply rooted. *Values* are very closely held; some values are so strongly felt that they may not be changed at all, or only slightly modified over long periods of time.

Understanding and then tapping into the system of expectations, beliefs and values of employees can be extremely difficult but very effective. One way is to empower; to provide space for self-development, creativity, and responsibility at work.

What, exactly, is motivation?

Motivation incites, directs and maintains behavior towards goals. Motivation and job performance are, of course, related; a motivated person who desires to do well at work is willing to expend effort to do so. Performance is the product of motivation and ability. Performance is moderated by situational constraints at work, which are factors that can stymie or enhance performance; examples are the climate (both psychologic and physical) and up-to-date equipment.

Motivating employees is an elusive yet, of course, highly desirable goal; understanding what motivates people can help us understand each other and can ultimately give us clues on how to make people as effective and as happy at work as possible. A number of theories that explain motivation exist, and a few of them are outlined below; they can be roughly divided into two categories. The first group focuses on the individual and his or her inherent traits; the second group places the environment at the forefront of motivation. Theories that focus on the individual include need-based theories and the expectancy theory. Environmentally based approaches, which assume that motivation is driven by external factors, include equity theory and reinforcement theory.

Behavior is apparently motivated by the urge to satisfy needs. These needs characterize and drive us, and we act in a continuous quest to satisfy them. Needs are defined here as requirements for survival and wellbeing and are not optional to an individual. Needs fall into two related categories: physical, those that are necessary to physiological survival and comfort, and psychological, which are required for a fully functioning consciousness.

Maslow (1943) developed the widely known *Need-Hierarchy* theory; later researchers then applied his work to organizational behavioral uses. Maslow and his disciples viewed motivation as a function of meeting an employee's needs, ranging from the basic physiological necessities (food, shelter, etc.) to the higher-order wants (self-esteem, self-actualization). He believed that individuals act in a never-ending mission to satisfy these needs, covering first the basic needs and then working their way up the hierarchy. He felt that lower-order needs were fulfilled before higher-order needs became desirable – the basic needs taking precedence over the higher needs. The hierarchy of needs he proposed has five levels: *physiological needs* such as the need for food and water are at the basis, followed by *safety needs* which focus on economic and physical security. The third rung revolves around *social needs* for belongingness and love, while the next level focuses on *esteem needs*, including self-confidence, recognition, appreciation, and respect. The highest-order needs are *self-actualization needs*; it is at this level that an individual achieves full potential and capability. Figure 2.2 depicts the model.

Alderfer (1972) took another need-based approach called the ERG theory. The acronym stands for three types of needs: *existence, relatedness,* and *growth*. Existence needs are material needs and include food, water, compensation, and working conditions; relatedness needs involve relationships and interactions with others, including family, friends, and colleagues; growth needs revolve around the desire for personal development and advancement.

There is some similarity between the two theories. Alderfer identified similar needs as Maslow did, but saw these needs arranged in a continuum as opposed to a hierarchy, which allows for movement back and

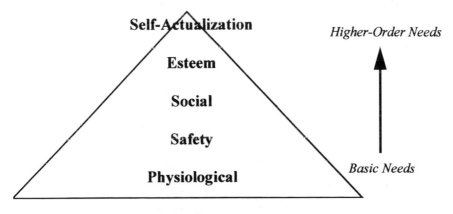

Figure 2.2   Maslow's motivation theory.

forth among the needs. He also hypothesized that if a person became frustrated in a fruitless attempt to satisfy a higher need, he or she would regress toward fulfilling lower needs instead; he called this "frustration regression".

Another approach is offered by the *expectancy* theory, which originated in the 1930s but was not applied to work motivation until the late 1950s; it then became quite prominent in the field of motivation research in the 1960s. The expectancy theory is a conglomerate of several researchers' ideas about motivation; it provides a clean, logical method to understand employees' motivation on the job and allows us to assess individuals' expenditure of effort at work.

Expectancy theory assumes that an individual's motivation (and resulting satisfaction) depends on the difference between what his or her environment offers versus what he or she expects. The theory can be expressed in form of an equation that outlines the factors underlying motivation. In its simplest form, it states that

> motivation is a function of
> {expectation that work will lead to performance} *times*
> {expectation that performance will lead to reward} *times*
> {value of reward}.

Here are some of the key tenets of this theory: it is a cognitive approach in which each person is assumed to be rational and knowledgeable about his or her desired rewards. The reward component in the above equation is generally something an organization can provide for its employees, such as pay, promotions, and formal recognition. It might be better described as an *outcome* because it can actually be something negative as well (getting fired, for example). It is also important to note that the value component (also called valence) is highly individual; it refers to how the employee rates the anticipated reward. Finally, the expectancy component is crucial; people must connect how hard they try and how well they perform.

A factory worker on an assembly line may have little incentive to increase her rate of production since the overall speed of the assembly line remains unchanged regardless of her efforts. On the other hand, a real estate salesman can conceivably earn more the more houses he handles, so he may be motivated to show as many homes as possible during a given workday.

Each of the above equation's components can help explain some aspect of motivation. We again take an example.

If an actor in a play performs well (expectation that work leads to performance) and receives critical acclaim for his acting in the piece (expectation that performance leads to reward), we may assume that he should be motivated. However, if money is the reward he valued rather than good reviews, and he does not receive adequate financial compensation for his success, he may lose his motivation in spite of the reward given (critical acclaim).

The expectancy theory has received prevailing support in the realm of organizational behavior and is widely applied today.

Reinforcement Theory – also called operant conditioning – originated with B. F. Skinner and his work with animals in the 1960s. His findings were applied to organizational behavior and used to describe motivation. The basic reinforcement model involves three key components: *stimulus*, *response*, and *reward*. The stimulus elicits a response; the reward is what is given to reinforce a desired response. Applying reinforcement theory to the workplace can be effective in motivating employees, but picking the appropriate rewards can be challenging. Individuals respond in different ways to different incentives.

Consider the dilemma faced by George, the manager of an upscale clothing boutique. Currently, his six salespeople are paid on a commission basis, where each sale is rewarded with pay for the salesperson making the sale. This is how George has run his store for the past 8 years, but now he worries that the competition among his salespeople is becoming unhealthy. Therefore he is considering revamping the existing commission-based pay system and implementing a straight-salary system, which, he hopes, would foster teamwork. At present, salespeople argue over who will work the most coveted and valuable (i.e., busiest) shifts; they also compete ferociously with each other on the sales floor to obtain customers. They often rush customers into purchases to make their commissions, and several customers have complained about the overly aggressive salespeople. Although George recognizes the benefits of a straight salary, he wonders about a potential drop-off in sales without the commission structure since the current system directly rewards sales. Three of the sales people favor a salary system to reduce inter-personal competition and infighting; the other three prefer a commission system because they feel it maximizes their income potential. The first three consider teamwork and a friendlier work environment a better reward than monetary compensation; the latter three find the opposite to be true.

Obviously, the reward must be meaningful and valuable to the employee but there are plenty of individual differences among employees, so any universal reward systems are inherently imperfect. This, in fact, is one of the primary limitations of the theory – although in principle it functions well, it frequently ignores individual differences in how rewards are valued.

The *Goal Setting* theory posits that people set targets and then purposefully pursue them; it is based on the premise that conscious ideas underlie our actions, and that what motivates us are the goals we have set (Locke and Latham, 1984). These goals both motivate us and direct our behavior; they help us decide how much effort to put into our work. To positively influence motivation and behavior on the job, however, employees must be aware of the goal, must know what actions are necessary to achieve it, and must recognize it as something desirable and attainable. Definite goals may foster higher levels of commitment on the part of the employee; and the more specific the goal, the more focused the efforts of the individual to attain it.

Summarizing what we know about motivation

In spite of how compelling these and other existing theories are, there is no clear-cut "winner" to the question of what motivates people. Instead, bits of all of the theories apply to different people, under different circumstances. Motivation is both intrinsic and extrinsic – factors within us and external to us drive our behavior. What most likely occurs is that we consider our own needs and wants and either consciously or subconsciously determine our goals; then we act to increase the chances of obtaining what we want. We do know from the above theories that we will strive first and foremost to fulfill basic survival needs like securing food and shelter, and that beyond this our needs are still very real but vary widely among individuals in terms of priority and strength. Other people influence us and shape our motivation and behavior because we are all to some degree social creatures. How hard we work is influenced by what our work will bring us: if we are rewarded in ways that are meaningful and valuable to us, and if we perceive a definite link between the strength of our efforts and performance, we will work strenuously to perform well and gain those desired rewards.

We should note here that what once motivated people has shifted, in many instances, in recent years. Pay and stability used to be considered the important motivators in a job; now, with an increasingly diverse, lifestyle-conscious workforce, there are many other non-financial ways to reward employees and keep them motivated and happy. Child care, on-site health clubs, flexible work hours, vacation time – these are just a few examples of the rewards that employees might seek; or perhaps an employee most values friendly coworkers in a nurturing corporate culture.

What, exactly, is job satisfaction?

Employees spend vast amounts of their time on the job, so it is not far-fetched to assume that job characteristics will influence job satisfaction, which will then affect productivity and performance at work. Job satisfaction is important to us as a society, as managers, and as employees for several reasons. First, as a society, we believe in a high quality work life as a goal; we feel that everyone has the right to fulfilling, rewarding jobs. Additionally, employers are increasingly interested in keeping employees happy because job satisfaction is associated with critical (and revenue-impacting) variables like turnover, absenteeism, and job performance. And employees who are happy on the job might well enjoy a higher quality of life overall, with fewer stress-related disorders and illnesses than those who are not. There does appear to be a causal link between job stress and physical disorders; stress is covered in further detail later in this chapter.

Shellenbarger (Wall Street Journal, 22 July 1998, page B1) cites research by retail giant Sears Roebuck that encompassed 800 of its stores to survey employees' levels of satisfaction. To measure satisfaction, Sears assessed attitudes on ten essential factors, including treatment by bosses, workload, etc. Sears found quite simply that a happy employee will stay with the company, provide better service to customers, and recommend company products and services to others. Specifically, Sears discovered that if employee attitudes on those ten essential factors rose by 5%, customer satisfaction (and, accordingly, company revenue) increased by 1.3%. The company is so convinced that employee job satisfaction contributes to customer satisfaction that it bases its executives' long-term bonuses in part on results of periodic employee satisfaction surveys.

Job satisfaction is defined as the extent to which a person derives pleasure from a job. It is distinct from morale, which refers to the collective spirit and overall goodwill of the larger group in which the person functions. Job satisfaction has been extensively researched and, accordingly, there are several theories that strive to explain what causes or prevents it. These approaches can be broadly categorized into need-based or value-based theories, social comparison theories, and job content and context theories.

According to need- or value-based theories, job satisfaction is an attitude that is determined within the individual. The theories in this general category postulate that every person has physical and physiological needs that he or she strives to fill in order to obtain satisfaction. A job defined as satisfying here would meet physical needs (such as food and shelter) through such aspects as appropriate income, and would meet psychological needs (such as self-esteem and intellectual stimulation) through growth opportunities and professional recognition. (We already discussed the most well-known need-based theories earlier when we reviewed motivation.) Needs not only motivate us, but they provide satisfaction when they are fulfilled.

Value-based theories recognize that although needs may be universal, people assign different weights and priorities to them. One person might value personal growth and recognition at work, for example, but care little about monetary rewards; this person would be satisfied even in a low-paying job as long as he or she were recognized for performance and given plenty of professional challenges. According to these theories, job satisfaction is present when a person's needs and values are met; if a person's needs or values change, however, dissatisfaction could result.

While the need-based theories explain a great deal about happiness at work, they neglect to take into account that people do not work in a social vacuum and that human nature compels us to compare ourselves with others. This brings us to the second group of theories, which center on *social comparison*. Here, researchers posit that people assess their own feelings of job satisfaction by observing others in similar positions, inferring their feelings about their jobs, and comparing themselves to the other people who work in similar capacities. Intuitively, it makes sense that social comparisons influence job satisfaction: our perception of others factors into our lives in infinite ways, and satisfaction at work is no exception.

Work conditions are the focus of the third group of theories that attempt to explain job satisfaction; Herzberg's two-factor theory is the best known and best researched of these (Figure 2.3). Here, job *content* factors, when present and positive, are associated with satisfaction, while job *context* factors, when negative, are associated with dissatisfaction. Herzberg and his colleagues isolated five content factors that they felt were satisfiers: *achievement, recognition*, the *work itself, responsibility*, and *advancement*. He proposed that a job featuring plenty of these factors would create job satisfaction, and that their absence would result in a neutral or indifferent employee. Further, he speculated that a different set of factors influenced dissatisfaction; these are the *context* factors mentioned above. Context (also called *hygiene*) factors include *company policies* and *administration, salary, supervision* and *management, interpersonal relationships*, and *physical conditions* at work. He proposed that a job without good context factors would make an employee dissatisfied, while the absence of context factors would lead to a neutral or indifferent employee. The upshot of Herzberg's theory is that a job

Figure 2.3 Herzberg's two-factor theory.

should ideally be designed with plenty of rewarding content factors (to ensure satisfaction) and positive context factors (to avoid dissatisfaction).

Ultimately, all the theories described above contribute to an overall understanding of what keeps employees happy on the job. Each theory supplies some degree of explanation for what contributes to job satisfaction. An integration of existing theories would likely provide the best approach to understanding satisfaction. Clearly, we all strive to fill basic needs – shelter, food, adequate clothing are mandatory to us and take precedence over higher-order needs. Once we have ensured the basics, we wish to be intellectually challenged, recognized and praised, promoted and rewarded. However, we all have different definitions of intellectual stimulation or of what constitutes a reward. Also, we will place different weights on what is important. Moreover, we are doubtlessly influenced by what we see around us – are our colleagues earning more than we are for doing the same work? Does our friend who works in jeans seem to be having more fun, or is it the one who wears designer suits and works in a window office? Do we like our peers at work? How supportive is our boss?

Summarizing what we know about job satisfaction

As a summary statement of the various factors underlying motivation and job satisfaction, we might do best to take a comment from the trenches. In a survey of the top US companies, one company that continually made the 100 Best Companies to Work For list (according to Candace Goforth, Knight Ridder/*Tribune*, 6 February 2000) offered these reasons for its success: "This is a positive place. People are friendly ...they feel challenged. They feel respected and valued. And they respond with loyalty." The company's philosophy included: "Treat others the way they want to be treated. Strive for mutual respect and for an atmosphere that makes people proud to work here. Provide career opportunities. Say thank you for a job well done."

Consider the case of Andrea and Betty, two former schoolmates who have forged very different careers for themselves. Both are extremely satisfied with their jobs. Andrea is a lawyer, on the proverbial fast track at a large inner-city law firm. She treasures the recognition that she gets for working long hours and winning the cases assigned to her, and her impressive record of promotions and steady career advancements help satisfy her desire for positive feedback and approval. Her employer's posh downtown law offices boast large, expensively furnished window offices; all partners receive an extensive support staff, generous expense accounts, and lofty annual salaries. In spite of the 80-hour plus workweeks, she feels that the fame and fortune of the job make it more than worthwhile. Her school chum Betty, on the other hand, works as a volunteer animal technician at a non-profit animal shelter in the same city. She works in a dilapidated building in which most of the space is devoted to operating rooms and kennels. Everyone – including the veterinarians – handle even routine tasks like answering phones, greeting visitors, and mopping up when needed. The vets and animal technicians are a close-knit, informal group. Betty's salary barely provides her with enough money to meet expenses, but she gleans rewards from colleagues' verbal praise and from the sight of recuperating animals. She appreciates the freedom of wearing shorts to work and rarely works more than 32 hours per week. This leaves her plenty of time for family and friends, her own pets, and her hobbies. Although both women have chosen very separate careers, with what Herzberg would call widely varying job content and context factors, both consider themselves extremely satisfied at work.

What, exactly, is job design?

We can also understand job satisfaction and employee motivation by looking at the way a job is designed. In the early 1900s, Frederick Taylor radically changed jobs by introducing specialization. He suggested that jobs be broken down into small tasks that were then to be standardized and divided among workers. This usually meant that a given worker did the same task repeatedly, resulting in extreme specialization – and, unfortunately, often extreme tedium. Although Taylor's job designs generally improved productivity itself, workers rebelling against the boredom, depersonalization, and routine of their repetitive jobs began showing up late, not appearing for work at all, and suffering from stress. (Highly repetitive work can lead to injury, as discussed with respect to keyboarding in Chapter 5.) In the mid-1900s, job enlargement (increasing tasks and variety) and job enrichment (increasing workers' participation and control) grew popular.

Several motivation and job satisfaction theories that we mentioned earlier can help to explain how job design influences behavior. The job-characteristics model (Hackman and Oldham, 1980) synthesizes much of the existing research on job design and is one of the most accepted and examined explanations of job enrichment. The job-characteristics model proposes that any job can be described via five core dimensions. These are (1) *skill variety*, the number of different talents and activities that the job requires; (2) *task identity*, the extent of work from beginning to end that the job involves; (3) *task significance*, the job's impact on others. These three factors make the job meaningful. The fourth factor, *autonomy*, describes the degree of independence in planning, controlling, and determining work procedures. This makes the person feel responsible. Finally, *feedback* provides information about one's effectiveness, how performance is evaluated and perceived.

All of these core job dimensions, when present, influence critical psychological states of the job holder through the experience of *meaningfulness of work, responsibility for work outcomes,* and *knowledge of results*. According to the theory, high levels of the three critical psychological states will lead to favorable work outcomes. These include high motivation, higher quality work, higher satisfaction, and lower absenteeism and turnover. Put more simply, the basic tenet of good job design is that employee happiness leads to quality of work life; one in which employees can fulfill their personal potentials.

Headlines on July 22, 1998:
"Best firms don't need employee loyalty" – *USA Today*, page 11A
"Companies Are Finding It Really Pays to Be Nice to Employees" – *The Wall Street Journal*, page B1

What, exactly, is power?

The word empowerment has appeared already in this chapter. It is a popular buzzword in organizations. In the intricate political landscape of an organization, power affects everyone. Power is a motivator; people tend to crave, use, and occasionally abuse their power, so it is helpful to understand where it originates.

There are five types of power: *coercive, reward, referent, legitimate,* and *expert.*

- *Coercive* power is based on fear; people who have this power are in the position to inflict some sort of punishment on others. At a restaurant, the shift manager might place a waiter who is on her bad side on a slow station with undesirable tables where tips will be low. An organization's coercive power includes its capacity to fire employees or dock pay.
- *Reward* power is just the opposite of this; it is based on the expectation of receiving praise. A teacher might design an especially detailed lesson plan in the hopes of receiving praise from the principal; the principal has reward power. A company's reward power lies in its ability to provide incentives such as raises, promotions, and paid vacations.
- *Referent* power exists when people admire a person regardless of his or her formal job title or status; generally, this person has particular charisma and an ability to attract and inspire followers. An example might be the secretary who, although officially called an administrative assistant, is widely known by company employees as the person who single-handedly keeps the office running. She knows where to find everything, when to reorder supplies, and with whom to talk to get things done.
- *Legitimate* power is due to the formal status held within the organization's hierarchy. A personal trainer at a gym has legitimate power over his clients – at least for the duration of their workout session. A corporate CEO makes the final decisions in corporate meetings because he/she has the authority and even the duty to do so.
- Finally, *expert* power comes from one's own skill or knowledge, regardless of formal position or job status. A low-level computer programmer may have extraordinary power in a company if he is the only one who knows how to revise the web site, for example, or to keep e-mail functioning. Another example is the management consultant brought in to evaluate the computer system of an organization; her perceived expertise in the computer field lends her this type of power.

Consider the case of an orthopedic surgeon in private practice who hires a medical assistant to help in her busy offices. The surgeon holds legitimate, coercive, and reward power over her assistant – she can fire him, give raises, and formally is the assistant's supervisor. Additionally, her expertise in the field of orthopedics gives her expert power over the assistant as well as referent power.

What, exactly, is stress?

Stress is a psychological state caused by environmental conditions that lead to a person's specific psychological, behavioral, and physical reactions. Stress does not discriminate; almost every worker in any job faces stress, and it occurs at all job levels. Because it is initially a psychological phenomenon, stress is closely tied to motivation, behavior, and job satisfaction (Kroemer et al., 2001).

Hans Selye was an Austrian citizen who emigrated to Canada in the 1930s where he worked and published on human reactions to overload, physical and psychological. He introduced the concept of *stressors* causing stress, even distress, if excessive. But, as he said in a presentation to the Human Factors Society in 1990, he simply did not have enough experience with the English language to realize that engineers consider stress to be the cause of strain – thus the confusing use of the term *stress* as either the cause or the result.

Stress as a psychological phenomenon involves interactions between an individual and the demands of the environment. The concept of stress has its foundations in research by Cannon, early in the last century, and especially by Selye (1978). They described what we now simply call stress as "general adaptation syndrome" (GAS), a pattern of physiological reactions to the environment, or the body's reaction to adverse environmental stimulation (Conway and Smith, 1997). Selye characterized GAS as a mobilization of energy resources that proceeds along three stages: *alarm, adaptation (resistance),* and *exhaustion*. In the alarm stage, the body mobilizes its resources to meet the assault of a stressor; these bodily resources include increased blood pressure and heart rate, elevated muscle tension, higher levels of hormone production, and the release of energy. During the adaptation stage, the body works hard to maintain homeostasis (physiological balance). This effort to normalize its systems strains the body. In the third and last phase, the person's biological integrity is endangered if the body's systems are overworked in their efforts to adapt – the normal stress has become an overwhelming distress. At some point, if overload is continued, the primary biological systems begin to fail, and serious physiological problems can occur. In short, Selye believed that response to a stressor led to heightened use of resources in order to either resist or adapt to the demand. This applies in a similar fashion to psychological loads and the person's responses as well.

When an individual is experiencing psychological, behavioral, and physiological effects of stress at work, actual changes in body chemistry occur, and these changes – which include heightened blood pressure, increased muscle tension, decreased immune system response, and hyper-ventilation – have been shown to lead to depression, headaches, and ulcers, to contribute to an increase in cardiovascular health problems, even to an increased risk of work-related musculoskeletal disorders (Carayon et al., 1999).

Selye hypothesized that stress followed a stimulus–response model, in which certain environmental demands (stressors) would invoke a related response. This early theory has been updated, however, to take into account our individual differences. In a given situation, individuals react differently to the same stressors, and any one person may even react differently at different times. A whole host of personal character-istics determine the physical and psychological effect that any given stressor will evoke in an individual. These include personality, current state of physical and mental health, skills and abilities, physical condi-tioning, prior experiences and learning, value systems, goals, and needs. In order to take these individual differences into account, a cognitive approach to understanding stress is now widely accepted. It refines the early straight stimulus–response paradigms.

Since individuals react in different ways to environmental demands, stressors vary in terms of the severity of their impact on people. We can, however, generalize and state that there are certain environmental demands that tend to exert the most significant effect on employees. Some of the biggest stressors include the following:

- *demanding jobs* with high work loads, pressure to perform, a too-fast work pace;
- *lack of control* over the process and work and lack of autonomy;
- *task difficulty*, with duties perceived as overly complex, or with conflicting demands;
- *overbearing responsibilities* for others, lack of social support, isola-tion;
- *monotony* and *underload* on the job, with overly routine, repetitive, and boring duties and little content variety;
- *poor supervisory practices*, including non-supportive superiors or incompetent management; and
- *technological problems*, like frequent computer malfunctions or equipment breakdowns.

Conway and Smith (1997) specifically researched the effects of compu-terization on job stress. Reviewing a number of studies done by various researchers, they concluded that psycho-biological mechanisms do exist that link psychological stress to heightened susceptibility to work-related musculoskeletal disorders (for example, in keyboarders) because stress can affect our hormonal, circulatory, and respiratory systems. This in turn exacerbates the impact of physical risk factors, such as improper workstations and excessively repetitive movements, discussed in later chapters in this book. One can argue, therefore, that proper ergonomic design of work tools and equipment is especially consequential in high-pressure jobs.

Reducing distress becomes an important goal for all levels of the organization, from the individual employee through supervisors to the company as a whole. The fundamental approach is to reduce the workload and other factors on the job that are unnecessarily stressful. Lower stress levels help employees enhance job satisfaction and maximize personal health; help supervisors keep their staff happy and healthy; and help companies achieve and maintain low healthcare costs and high productivity. A number of researchers have examined approaches for managing stress, as reviewed by Muchinsky (2000) who listed six major coping mechanisms. These were:

1. setting priorities, deciding what is really important and what is not;
2. utilizing home resources, like talking about feelings with family members or friends;
3. recovering and dealing with the problem by moving on to a new activity;
4. distracting oneself with an unrelated activity or by taking the day off;
5. passive toleration by giving up and accepting the situation (not so good);
6. releasing emotions by taking feelings out on others (also not usually desirable).

Unfortunately, quite a few people deal with stress by smoking, drinking, turning to drugs, and withdrawing from others.

In recognition of the many negative effects stress can exert on employees, some companies have created and implemented formal stress-reduction initiatives and programs. These include time management seminars, instruction on relaxation techniques, on-site wellness facilities that include exercise equipment and yoga classes, on-staff psychologists or counselors available to employees, biofeedback seminars, and more.

What, exactly, is communication?

One activity that pervades and influences – directly or indirectly – all of the topics we have discussed above is *communication*. It is the exchange of information between two (or more) persons and the inference of its meaning. Individuals understand and establish their roles in an organization through many modes of communication: how they provide and receive feedback, how they make decisions, how they state and pursue goals. Functions of communication include establishing control, disseminating information, enhancing motivation, and expressing feelings. Muchinsky points out that communication is the lifeblood of an organization – and that failure to communicate is its Achilles heel. Communication occurs throughout the organization and the environment in which the company operates – interpersonally (among employees), intra-organizationally (among the groups and departments of the company), and inter-organizationally (between organizations). If you have ever perused the job classified ads, you will have noticed how many companies request "good oral and written communication skills" in their job specifications: the value of positive communication cannot be overstated.

Interacting with others

Surviving and, ideally, thriving in the intricate political structure of an organization virtually mandates getting along with people. The people with whom you most need to interact effectively are your subordinates and your supervisor. Managing the relationship upward is as important as managing the relationship downward. In order to manage your boss, you must understand the boss – including his or her context, goals, the pressures she or he faces, strengths, weaknesses, and preferred work styles – and then you must understand the same things about yourself. This done, you can map out, develop, and maintain a relationship that fits both your needs and styles and meets both of your expectations. This relationship should be based on dependability and honesty and selective use of your boss's time. Keep supervisors informed – they rarely appreciate surprises.

The steps above may seem somewhat idealistic – how can we truly understand the stated and unstated goals of our bosses? Yet even just trying to understand them will bring us closer to effectively dealing with colleagues. If you know, for example, that one of your supervisor's goals is to lose ten pounds by Christmas by working out at the gym every night, you will not schedule meetings in the late afternoon. If you realize that your boss's style is formal and old-school, you will draft formal memos to schedule staff meetings rather than leave voice messages or e-mails. These basic principles can be similarly applied to dealing with subordinates – understand them, understand yourself, and base your relationship with them on these assessments.

One tool that will help you to interact with supervisors, subordinates, and colleagues alike is a technique called active listening. Many times, when we assume that we are listening, we truly are not; instead, we might be silently disagreeing with the speaker (*he does not know what he is talking about*), formulating our response to the person talking (*well, you could have sent a letter*) or thinking of something else and tuning out entirely (*I wonder what I should have for lunch*). Active listening helps circumvent this problem by forcing us to really hear what the other person is saying. This lets us get a valid perception of what the speaker is communicating – and gives the speaker the satisfaction of truly being heard. When you practice active listening, you pay close attention to what the person is saying. You respond to the information the speaker gives you without leading and without giving advice, and you identify the speaker's feelings along with the content of what was said by absorbing the information and repeating portions back to him or her to verify the information. Importantly, an active listener gives control of the conversation to the other party.

Performance appraisals

Performance appraisals are formal means of communicating among managers and employees. When done properly and at the right time, performance appraisals can be very useful in achieving *organizational/ administrative goals*, providing *feedback and evaluation*, and for *coaching and development*. Organizational or administrative goals pertain to personnel actions such as placement, promotions, and pay (and, when need be, provide the necessary documentation for firing). Feedback and evaluation refer to the employee's performance, including strengths and weaknesses of the employee's work. Coaching and development relate to the critical final goal of the appraisal: how to improve – rather than punish – the employee's efforts. Working together, the manager and the subordinate should agree on specific goals and timetables for improvement; this provides a valuable opportunity for encouragement and career coaching.

While they are very valuable in the theoretical sense, in practice, performance appraisals are often misused, mismanaged, and widely feared. People resist formal performance appraisals for a number of reasons. Managers may dislike giving evaluations because they are uncomfortable criticizing subordinates and feel that they lack the skills needed to evaluate employees. Subordinates often become defensive and anxious during their appraisals. Accordingly, the task is often delayed until the appraisal has effectively lost its usefulness. Management plays a vital role in making the appraisal process succeed by tying the process into the organization's overall role. Appraisals must have tangible results; for example, if promotions and raises are given based on performance, the appraisal process should be the tool for measuring and communicating performance. Put differently, there should be a high correlation between the thoroughness of the appraisal process and how extensively the resulting information is used.

Performance appraisal systems often include other types of evaluations in addition to the downward (boss-to-subordinate) evaluation. These include self-assessments, peer assessments, and upward or management evaluations (employee-to-supervisor); all provide important data for use in the evaluation.

- *Self-assessments* – people are asked to evaluate themselves and outline their strengths and weaknesses. If nothing else, self-assessments provide a good springboard for dialogue between the employee and the supervisor and may help reduce some of the defensiveness that often creeps in when the employee feels lectured to, given the propensity for a downward evaluation to be fairly one-sided.
- *Peer assessments* – members of a group (a department, account, or team) are asked to evaluate their fellow colleagues. In certain service industries, client counterparts might also be polled concerning an employee's performance.
- *Upward evaluations* – managers who oversee employees should be evaluated on their supervisory ability as part of their overall appraisal. Who best to provide input on supervisory skills than the employees themselves?

A significant part of the appraisal process involves gathering the information and reaching a conclusion regarding the employee's performance. Once the appraisal itself is completed, the appraisal interview takes place; this is the critical concluding link of the appraisal process in which supervisor and employee meet to review and discuss the evaluation. Not surprisingly, both supervisor and employee are generally uneasy and anxious about the actual interview. Few people look forward to the confrontational nature of an appraisal interview; both parties often feel nervous, tense and defensive.

Several factors play a role in how comfortable this meeting will be, including the method of delivery (a supportive supervisor is more effective than one who is threatening), the existing relationship between the supervisor and the employee, and the non-verbal cues that are projected during the interview. In general, however, an interactive mode of communication seems to improve the tenor and outcome of the interview. In other words, giving employees substantial participation in appraisal interviews and conducting two-way discussions increases satisfaction with the interview and sparks motivation to improve subsequent job performance. In addition to a review of past performance and discussion of strengths and weaknesses, the appraisal interview should involve setting job-related plans and goals for the future (an action plan for the employee). This action plan should be collaborative – based on the views of both parties – and realistic; otherwise, it may become meaningless or even detrimental. (If the employees do not buy into the action plan, or find the goals set for them impossible to achieve, they may actually lose motivation since their job prospects appear dim.) The action plan is an instrumental tool in evaluating performance during the *next* evaluation.

Balance in work and personal life

"A rested employee is essential to a company's business." This statement, from a *Wall Street Journal* article (Joann Lublin, July 6, 2000, page B1), is one many of us who have worked in corporate America never felt would be uttered or embraced. However, companies in the U.S.A. are beginning to recognize the negative consequences of overworked and overwhelmed workers. Employees need time off-the-job and vacations to develop a balance between work and private life. Technology seems to simplify our work – and makes it more complex at the same time. Consider time-saving devices like laptop and handheld computers, pagers, and cell-phones – they have doubtlessly helped us streamline our work, yet they enable (even force) us to be connected to our jobs at all times.

Companies' interest in employee welfare, for example in terms of balancing work and leisure, is not simply a touching demonstration of altruism. Instead, as mentioned earlier in this chapter, a company's value is determined in large part by its soft assets, its workforce. Some share-holders go as far as to gage employer respect for work–life balance as a basic indicator of organizational health and soundness (Sue Shellenbarger, *Wall Street Journal*, July 22, 1998, page B1). They recognize that frustrated, overtaxed employees who do not have a life outside work are less effective in the longer term.

To help employees balance their time, some companies restrict week-end and vacation work, while others are implementing sabbaticals and offering personal time. The article by Joann Lublin (*Wall Street Journal*, July 6, 2000, page B1) points to a dot-com company in Manhattan that lets its staffers take Wednesday mornings off to do personal errands (or just sleep in). Radio Shack Corporation's Guilt-Free Vacation policy instructs new employees not to call the office while on vacation. Other companies are strictly limiting meeting times to keep employees from spending so much time in meetings that they must devote weeknights and weekends to catching up on their regular work. Some companies have implemented meeting-free days.

Even high-end management consulting firms, notorious for extraordinarily rigorous work schedules that often leave their employees overtaxed, are looking for ways to reduce the workload and let their employees have a life outside of the work. Consulting staff often work on client engagements that require frequent travel, and in the past, consultants would travel to the assignments on Sundays to spend the following week at client offices. Some companies have eliminated weekend business trips for their consultants; now, they do not leave for client assignments until Monday, giving them their weekends to themselves.

ERGONOMIC RECOMMENDATIONS

Both managers and employees alike have many factors to consider when they decide to join or leave an organization, hire an employee, or seek ways to make their work lives more pleasurable. While each person and employer and job is different, a group of attributes that are culled from the literature can likely satisfy people in their jobs and make them successful and productive. Table 2.1 presents an overview of factors that are generally important.

The most important considerations for employees and managers are given below, with separate recommendations for each, along with a further list of recommendations that apply to all of us:

Ergonomic recommendations for employees

Employees should make every attempt to understand their environment when they consider accepting work in any organization. Environment includes both the organization itself and the individuals within it.

What to do if you feel miserable

- Talk with your supervisor in a non-confrontational tone about existing problems. It is important that you propose solutions.
- Check your employment contract to determine your duties and rights at work.

Table 2.1   Major determinants of job satisfaction and work performance

---

Person
Personality
  Self-esteem
  Ability to cope with stress
  Health and age
  Satisfaction with life
  Skill in working with others
Control over one's own work
Achieved status and recognition

Organization
Reward and recognition system
Perceived quality of supervision
Decentralized power structure
Stimulating work and social relations
Job security
Pleasant working conditions

This makes you feel good:

- Examine the components of a company – like its structure, strategy, and policies – and determine if they agree with your own personality, your personal goals and beliefs.
- Pay particular attention to the culture of the organization; it will greatly influence you at work. Choose a company that fits your own personal style as much as possible.
- Job satisfaction is crucial; it affects our lives outside of work, our mental health, and our physical health as well. Consider both internal and extrinsic factors:

  - Examine your own needs and be as realistic as possible about what they are. Does the job meet your physical needs, like sufficient pay and fringe benefits? What about your emotional needs, like recognition and a sense of achievement?
  - Evaluate your work conditions – the office, the dress code, and so forth. Make sure they are appropriate for and pleasing to you.
  - Spend some time with your prospective colleagues and supervisors and determine how you feel about them since you will be with them many hours at work.
  - Remember that we are all different; what is important to you may not be important to someone else. That said, take a look at how you feel about your job compared to how others in similar positions feel about theirs. This may become increasingly important to you since social comparisons are part of human nature.
  - Take advantage of any programs offered by your company to increase your skills.

- Inevitably, you will be exposed to stress. Try to find healthy ways to cope with it if you find it becoming a problem:

  - If you are stressed, talk about your feelings to empathetic family members or to trained counselors; take time off; prioritize your job duties to feel less overwhelmed; enlist the help of others if you are overworked.
  - Take advantage of any stress-reduction programs your company offers.

- Most of us benefit from having – and protecting – a balance between work and our personal lives:

  - Technology can make our lives easier, but might be keeping us connected to work day and night. Make sure that laptop and

handheld computers, pagers, and cell phones are contributing
to your life, not just extending your workload.
– Employers should respect your personal life. Barring unusual
circumstances, you should be entitled to uninterrupted vacations, time off, and weekends.

- Consult your employee handbook or company policies and procedures manual if you have questions or concerns about your job duties,
  vacation and days-off allotment.
- Talk with your organization's union steward and social consultant.
- If there is no solution to a problem that bothers you so much that you
  do not want to work at the organization any more, then look for
  another position.

Ergonomic recommendations for managers

Poor motivation and lack of job satisfaction can lead to poor performance, absenteeism, and turnover. Not only will satisfied, motivated
employees be more productive, but they are also less likely to suffer
unduly from stress. Here are some pointers to enhance your employees'
lives:

This makes people feel good:

- Make sure your employee's necessary physical and psychological needs are fulfilled – physical needs focus on existence, health
  and comfort; psychological needs are those required for being at
  ease and feeling well.
- Remember that people place different weights and priorities on
  needs, and pay attention to what is important to different
  employees.
- Work conditions – especially policies and procedures, administration and supervision – play major roles in employee satisfaction; find out what conditions are important to your staff and try
  to provide them.
- We work for rewards – but since individuals feel rewarded by
  different things, offer rewards that are meaningful and valuable
  to them. This means finding out what people want before struc-
  turing rewards and incentive programs.
- Make sure there is a definite link between performance and
  rewards; one important tool for measuring performance and
  linking it to rewards is the performance appraisal.

- Human nature compels us to compare ourselves with others, so keep in mind that employees' motivation and satisfaction are also influenced by their comparison of themselves and of their jobs with others. This statement has many implications; for example, if salaries of employees doing similar jobs vary, the figures should be kept confidential.
- Provide opportunities for growth and learning; this enhances employee self-esteem and reduces job monotony. This includes offering on-going training programs.
- Increase the amount of thinking and decision-making in an employee's job; this means letting employees self manage when possible or feasible, letting employees get involved in the overall process of their jobs, and letting them contribute to corporate decisions.
- Make sure performance standards or work output requirements are reasonable; help protect employees from excessive work loads.
- Give employees plenty of breaks and vacations; protect their non-work lives and down-time.
- Provide opportunities for socializing.
- Make sure work stations are comfortable, well equipped, and in proper ergonomic conditions. (This is covered in the following chapters.)

Stress is a fact of work life and can be very detrimental to the health and happiness of your employees. To combat stress:

- Provide a supportive environment for employees, paying particular attention to direct supervisors – they should be particularly supportive.
- Consider offering stress-reduction programs if feasible; these may include time management seminars, wellness programs, relaxation training, etc.

Ergonomic recommendations for employers and employees

This makes you feel good:

- Understand your bosses and their context, including their goals, the pressures they face, their strengths and weaknesses, and their work styles.
- Understand yourself, assessing your own strengths and weaknesses and your personal style.
- Incorporate the first two steps and develop and maintain a rela-

tionship that fits both your needs and style, meets your expectations, and is based on dependability and honesty.

- Practice active listening when interacting with colleagues, which entails playing close attention to what your conversation partner is saying, responding to the information he/she gives you without leading or giving advice, and identifying your partner's feelings and meaning by absorbing comments and repeating portions to verify the information.
- If you are having trouble dealing with difficult persons, it is often best to go directly to them and discuss the problem calmly rather than going around them or complaining to other colleagues.

References

Alderfer, C. P. (1972). *Existence, Relatedness, and Growth: Human Needs in Organizational Settings*. New York: Free Press.

Carayon, P., Haims, M.C. and Smith, M.J. (1999). Work Organization, Job Stress, and Work-Related Musculoskeletal Disorders. *Human Factors 41*, 644–663.

Conway, F. T. and Smith, M. J. (1997), Psychosocial Aspects of Computerized Office Work, in M. Helander, T. K. Landauer, P. Prabhu (Eds.), *Handbook of Human-Computer Interaction*. Amsterdam: Elsevier, 1497–1517.

Greenberg, J. and Baron, R. A. (2000). *Behavior in Organizations* (7th ed.). Boston, MA: Allyn and Bacon.

Hackman, J. R. and Oldham, G. R. (1980). *Work Redesign*. Reading, MA: Addison-Wesley.

Herzberg, F. (1966). *Work and the Nature of Man*. Cleveland, OH: World Publishing.

Hendrick, H. W. and Kleiner. B. M. (2001). *Macro-ergonomics: An Introduction to Work System Analysis and Design*. Santa Monica, CA: Human Factors and Ergonomics Society.

Kleiner, B. M. and Drury, C. G. (1999). Large-Scale Regional Economic Development: Macro-ergonomics in Theory and Practice. *Human Factors and Ergonomics in Manufacturing 9*, 151–163.

Kroemer, K. H. E., Kroemer, H. B. and Kroemer-Elbert, K. E. (2001). *Ergonomics: How to Design for Ease and Efficiency* (2nd ed.). Upper Saddle River, NJ: Prentice Hall.

Landy, F. J. (1989). *Psychology of Work Behavior* (4th ed.). Pacific Grove, CA: Brooks/Cole.

Locke. E. A. and Latham, G. P. (1984). *Goal Setting: A Motivational Technique That Works*. Englewood Cliffs, NJ: Prentice Hall.

Maslow, A. H. (1943). A Theory of Motivation. *Psychological Review 50*, 370–396.

Muchinsky, P.M. (2000). *Psychology Applied to Work* (6th ed.). Belmont, CA: Brooks/Cole.

Selye, H. (1978). *The Stress of Life* (rev. ed.). New York: McGraw-Hill.

3   Office design

Overview

Many of us who are working in an office will spend about one-third to one-half of our waking hours in that office, perhaps including weekends. So, our office's design should be done in such a way that we are happy with it. The interior design of an office can affect our quality of life; additionally, there is a great deal of anecdotal evidence that it can also appreciably influence employee productivity. Consequently, an organization can help increase its chances of economic success by designing its office space so that it functions efficiently and effectively.

Large or small offices?

Office designs can vary from an open plan to closed, walled-in individual offices. Open plans include the stereotypical paperwork factory set up in a huge open room with straight rows and columns of desks and chairs. Figure 3.1 depicts an example of such an old-fashioned layout with rather modern computers on the desks.

Frequently, large spaces are subdivided by low partitions into smaller semi-private cubicles, which have often been featured as the object of many jibes in the razor-sharp Dilbert cartoon series. In spite of their popularity as the target of jokes, however, if done appropriately, partitioned cubicles can indeed help provide some privacy to employees.

At the other extreme of office design are the separate, walled-in offices used by just one person or shared by two or three employees. In reality, many offices are hybrid designs; they have some areas that feature open plans, while other departments or floors within the company may have individual, separately contained spaces.

The appropriateness of a given design depends on two main factors: the kind of work performed in the office, and the types of personal demands and requirements that the individual users themselves have. There are advantages and disadvantages to all types of office design.

Figure 3.1   Seemingly an antiquated mass production factory with the same identical workstations set up in straight columns and rows – but note the
computers on the desks.

Proponents of open offices praise cost effectiveness in terms of
construction cost and in terms of expenses for daily air conditioning,
or heating, cooling and ventilating the space. They also like the uninterrupted lines of communication among the employees and the ease of
collaborating with co-workers. Those who prefer closed-type designs
advocate the privacy and individuality of personal offices, recognize
the value of separate spaces for concentration-intensive tasks, and
appreciate fewer disruptive factors like extraneous noise.

The purposes of office design

Whatever the office size, proper space design must minimize the adverse
effects of environmental conditions such as noise and bad air; it must
provide effective lighting as well as appropriate interactions among
personnel. The office is a place to work and perform. In many cases,
offices are designed also to convey an image of the organization, to
express the personality of the company, and to offer pleasing esthetics.
The office's appearance contributes to attracting new employees, and

also helps retain workers. All of these factors work together to increase employees' job satisfaction and performance.

The process of office design

The process of office design – how to go about it – involves several important steps. The first step is to analyze the needs of the people who will inhabit the office, including the tasks they will perform, the machinery and equipment they will use, and their preferences and work styles. Based on such analysis, the ergonomist/designer should formulate specific statements of these functional requirements; they guide the actual design. The next step is to identify a range of design options so that practical solutions can be chosen; this reflects the need to compromise in the real world since the conceptual ideal is not always possible. Then, these options are evaluated and final designs are selected. Finally, the chosen design is implemented and put to use, with plenty of training and support provided for the employees. Note that these steps also apply, in principle, if an existing office is remodeled or otherwise updated. Employees should be involved in *all* phases of the process so that their needs are truly met and they do not feel manipulated or ignored.

Individuality and flexibility

A pervasive trend in office space design today is the consideration of individuality and personal direction, made possible especially by advances in technology. Current engineering know-how allows us to design almost any environment we wish. Yesterday's technology tied us to one office, and often to one place therein, because we were bound by wired phones and computers, cabled machines, and stationary equipment; now, many of us can be flexible in location, work schedules and habits.

You may skip the following part and go directly to the Ergonomic Design Recommendations at the end of this chapter – or you can get detailed background information by reading the following text.

A short history of the architecture of offices and office buildings

The building of the Uffici (Offices) in Florence, Italy, finished in 1581, was in U-form around an open court so that every room had a window to the outside to provide natural lighting and ventilation. Into the 1900s, large office buildings, and the arrangement of the offices within, still followed that example.

The Milam building in San Antonio, TX, was the first to be fully air-conditioned in 1936. This technology allowed huge, city-block wide edifices to situate many rooms away from outside walls (Arnold, 1999).

In France, at about the same time, Le Corbusier designed large office buildings without the customary Beaux Arts style of external ornamentation but featuring, instead, large glazed surfaces of metal-framed windows; this became known as the International Style. These buildings boasted Gustave Lyon's "regulated air" (l'air ponctuel) together with Le Corbusier's "neutralizing walls" (murs neutralisants) with air at a constant 18°C circulating between double window panes. The first such edifice was planned for Russia's Moscow, but it was not completed; instead, the Cité de Refuge in Paris, finished in 1933, had these features.

In 1932, the PSFS building in Philadelphia was erected in the new style, and fully air-conditioned, which the contemporary Empire State and RCA buildings in New York were not. The one edifice that really utilized all the new technology was Frank L. Wright's Johnson's Wax Administration Building, completed in 1939. It was entirely sealed and fully air-conditioned. It had clerestory windows constructed from bundles of glass tubes that provided diffuse light to interiors rooms (see Chapter 7).

After the Second World War, air conditioning of large and small office buildings became commonplace in North America. In the course of this development, many concerns grew as well: about the effects on health and performance of lacking natural light and the attempts to imitate it by luminaires (Chapter 7); about acceptable indoor air quality and about the work performance in offices that are too cool or too warm (Chapter 9); and about the results of putting many persons together in large rooms in terms of supervision, performance, acceptance, behavior, and satisfaction, as discussed in the foregoing chapter.

What, exactly, are the purposes of office space design?

Proper ergonomic design of the office space takes a focused systematic approach in which use- and user-centered requirements are utterly fundamental to the outlay of the space. In the ideal world, we would approach office space design scientifically, identifying and understanding the needs, capabilities, and limitations of the people who will occupy the space and then applying this knowledge in a systematic way. Use-centered requirements include the tasks or functions to be performed in the space and the equipment and machinery needs associated with these tasks. User-centered requirements concern the comfort needs and preferences of the workers, esthetic appeal and, of course, safety considerations. In the real world, however, we must also take into account budget constraints, size limitations, hierarchies within organizations, and time schedules. In most cases, after all, organizations cannot build anew but instead have finite square footages available to them. Moreover, that available space can be re-configured only under given restrictions, office work must continue in the meantime, and there are deadlines for office and personnel moves. Acknowledging the real versus the ideal world, the logic and procedure of responsible office design apply to creating new spaces as well as to re-arranging existing ones.

The office environment

Offices inherently expose people to environmental conditions that may affect their health, comfort, and their ability to perform. It is the task of office design to provide the best possible conditions and, certainly, to minimize any unavoidable adverse effects of environmental factors. Control of lighting (Chapter 7 in this book), sound (Chapter 8), and climate (Chapter 9) are the most important environmental conditions influencing how office inhabitants feel and perform.

Lighting, which is covered extensively in Chapter 7, strongly affects us in our work environments. Human factor engineers can set up lighting in the office so that visual work tasks are easy to do. The lighting engineer's design goal matches our own: we want an office environment that allows us to see clearly and vividly what we want to see, prevents glare and annoying bright spots in our visual field, and pleases us in terms of contrast and colors. Task lighting – a lamp or light source attached to and controllable at the workstation – is also becoming much more widely adopted. Where we once saw predominantly high levels of overhead lighting, light sources concentrated on the task are increasingly popular. Task lighting saves energy, is more appealing aesthetically, and gives us control over our own light source, which leads to additional individual comfort and satisfaction. A number of companies also recognize the value of natural light and are including glass and windows to add light and an open, more spacious feel to the work place.

Sound and its evil twin, noise, pervade our space in a myriad of forms – examples include ringing, beeping, chiming phones; churning, chirping fax machines, clattering printers and copiers; colleagues' conversations; a co-worker's radio; even external commotion from raging traffic, nearby trains or boisterous pedestrians passing by. We are accustomed to a certain background sound level, often generated by the climate control system, which helps to mask some disruptive noises. Even *lack* of sound can be disruptive – it can actually be too quiet (see Chapter 8).

Research and personal experiences suggest that noise usually is the most intrusive of the environmental factors influencing people in the office. Accordingly, acoustics should be prominently considered when planning office spaces. Acoustical requirements in offices consist of privacy of speech at both low and normal voice levels plus the ability to perform normal office work with minimal disruption or distraction from standard office sounds like another worker's voice. To help reduce noise, we can include tools like acoustically sound-absorbing screens, carpets, draperies, and outer shells of filing cabinets in our offices.

Air ventilation, temperature, humidity (and, relatedly, smells and other contaminants) also influence our comfort and, in their extremes, our health at work. Consequently, they too have an impact on our productivity, performance, and willingness to spend long hours in the office. In our work environment, the climate (the term is used here in its physical meaning) should be neither too hot nor too cold; neither too damp nor too dry, we also want fresh rather than stale air but the air flow should not generate a strong draft (see Chapter 9 for more details). The latest office spaces offer unprecedented levels of employee comfort; some even offer individual climate control at each workstation.

Looks good, feels good

Another function of office space design is to attract good employees, keep them on board, and increase their job satisfaction. Indeed, current initiatives in office design are centered on retaining employees and encouraging them to expand their office hours beyond the usual working hours. Without a doubt, altruism once again is not at the heart of the matter; instead, organizations are hoping for higher profit via increased productivity. With people spending long hours at work, one of the ways to keep them productive is to make them comfortable and happy with their office space. Job satisfaction, after all, affects productivity and performance, and work conditions have been shown to influence the level of an employee's satisfaction on the job (see Chapter 2 for additional information on the link between work conditions and job satisfaction).

An appealing office design may help attract prospective employees, and in the current tight labor market, qualified job seekers can look forward to a number of amenities. In fact, many organizations presently involved in interior build-outs are insisting on innovative options to draw personnel. Popular types of amenities focus on convenience and on creating a home-like atmosphere at the office. These features go well beyond the typical employee cafeteria and may include free brand-name coffee and juice bars, fully stocked larders, fitness centers that feature posh locker rooms with laundry facilities and on-site dry cleaning, child care, even on-site massages and hair cuts. All of these make it easier for people to be at the office. Making the office like an extension of the home and keeping it as convenient as possible become especially important when employees' duties require extended hours or overtime.

The design of an office, and how it is run, express the corporate character or personality of an organization. In a sense, the office serves as a company's visible motif. As an illustration, consider a fledgling company

hoping to attract young, hip computer-philes accustomed to unstructured environments; quite likely, the company's floor space will reflect
its upstart, unfettered nature.

Oxygen, a women-oriented cable channel with a web site network, took
over the top floors of an old factory in 2000; the office featured a large
range of activities and tasks. Bubbling fish tanks, interior balconies, and
rippling scrims decorated the office. As one of the company's co-founders
pointed out, "we wanted a cool office that was fun to work in so that we'd
attract digital kids." (Iovine, 2000).

Imagine that you are running a mid-sized consumer goods company. If
you are hiring an agency to do your company's advertising campaign,
and you are seeking a highly creative Cannes-contender for your television ads, you might expect to find stylized, irreverent offices stocked with
pool tables and cappuccino machines when you visit the agency. On the
other hand, if you are hiring a CPA firm to handle corporate taxes and
the annual audit, you might prefer retaining a company with more traditional furnishings and office spaces.

Range of office designs

Office designs can vary from large to small, from open plan to walled in
offices for one person. The appropriateness of a given design depends on
two main factors: what kind of work will be performed in the office, and
what types of personal demands and requirements the individual users
themselves have. Open-plan offices are ones in which people share space
and, often, office equipment; these types of office designs vary in their
degree of openness. They might be truly open, with no dividers or separa-
tors to section off desks or work areas, and frequently the workstations
are all the same, arranged in strict rows and columns, as shown earlier in
Figure 3.1. One variation is to allow different workstations in the room.
These may be arranged irregularly; tall dividers can outline clusters of
workstations or generate single stations. Another variation is to retain
the wide open general space but to subdivide it into smaller sections by
dividers or panels that may just be shoulder-high; this generates the
proverbial cubicles. Raymond and Cunliffe (2000) show many examples
of modern office layouts.

Some sections may be truly separated with solid floor-to-ceiling sound-
proof walls. Of course, smaller spaces for a smaller number of persons
have always been used, as have the single person offices. And as most of
us have experienced, being assigned an individual office even if it does
not have a window, or, better, a single-person office with a window, even

better yet with corner windows, has long been a sign of rank and importance and power.

The office landscape is based on an idea that took hold in Germany, and then elsewhere, in the 1960s. Its original objective was to ensure efficiency in the use of personnel and space (Kleeman, 1991). The basic concept of office landscape was to maximize the use of space and facilitate contact among the workers who need to interact in order to conduct business. The landscaping aspect refers to the arrangement of furniture, office machines, flowers and bushes and small trees to create the appearance of an irregular, natural terrain similar to what one would expect in a park or large garden. This often resulted in rather pleasant office environments that, on one hand, took up much space but, on the other hand, could also accommodate change and allow for expansion as a company evolved and grew.

There are pros and cons to all types of office design. Advantages of open spaces are that they are generally less expensive than separate closed offices in terms of building cost, tax write-offs, and utility usage for lighting and climate control. Moreover, open-plan spaces often facilitate communication and collaboration among workers. Additionally, proponents of open spaces praise the lack of hierarchy that individual offices may encourage – without the corner office suites that we may find in more traditional office spaces, there is less of a division between the executives and the staff.

Disadvantages of open designs include disruptive noise – such as speech and equipment sounds – which may reduce performance and job satisfaction. The lack of privacy of an open design may also be an issue. This is especially true for offices in which business that requires the exchange of confidential information is conducted. The fact is, some organizations – or even departments within a given organization – are not suited to open landscape designs because they inherently require more privacy.

Consider the human resources department of a multi-national financial services firm employing thousands of workers. The ten-person human resource staff is responsible for all personnel duties like hiring, firing, salary negotiations, benefits administration, etc., and so all ten members of the staff have individual offices with ceiling-to-floor walls. Given their daily tasks, which often require meetings and interviews with employees where confidential information is discussed, a high degree of privacy is necessary. Another example would be a research-intensive firm in which solitary invention dominates; here again employees probably benefit from private offices in which they can conduct their highly focused research tasks without interruption.

Problems with open plan offices can also actually extend to physical health: for example, upper-respiratory tract infections and headaches are more common in open offices when the climate-control system does not function adequately.

One popular aspect of landscaped offices was the use of plants: partly for their pleasing appearance, partly for their assumed abilities to prevent noise propagation and renew the air. Unfortunately, they do next to nothing in absorbing sound because they are not dense enough. And they are also not effective in absorbing volatile compounds or other contaminants in the air (Hedge, 2000). Still, they can be pleasing to the eye – esthetics *are* important.

Why must an office look like a paperwork factory? The corporate culture can produce rather unusual workspaces. Raymond and Cunliffe (2000) present the example of an old film studio that now houses 26 workstations for 75 people. With its attractive, irregular layout it looks more like a club than an office; it has a billiard table, a cooking facility and a miniature forest.

In summary, then, the appropriateness of an office layout depends on the type of work performed, including careful evaluation of the degree of interaction required and the amount of privacy needed. In addition to analyzing the nature of the business, we should also consider individual characteristics and preferences; some people are better able to concentrate on the task at hand even with disruptions, while others can concentrate only when they have complete privacy. Some people are inspired when they share a large table with half a dozen other people, when nobody can hide behind a closed door, and when they can simply take their laptop computers to any desk or chair within a large communal space. Others are uncomfortable without the security offered by office walls and feel oppressed by the forced communality of an open area; they prefer the privacy of their own personal space. Some find open spaces liberating; others find them limiting.

Many offices, of course, house a mixture of office designs – wide open spaces here, sections with shoulder-high cubicle dividers there, closed-off individual offices along the perimeter of the room. In many businesses, tasks that require concentration are interspersed with those that require interaction. The majority of us perform a number of tasks during the typical day, and we generally shift between individual work and work with others that demands team effort.

The process of office design

Office design is very specific. There is no one absolutely right design that fits every organization. Yet, a certain step-by-step procedure is common to all well-planned designs. The success of the design project relies on first gathering all needed knowledge about the work to be done in the

Office space can affect company culture

A large Chicago-based professional services company moved from an older office tower downtown into a newly renovated building a few blocks away. The company, a venerable institution with a rich 60-year history, employed thousands of people; one of the reasons cited by many for its enduring success and impressive growth was its extraordinary corporate culture. Employees often described themselves as part of a family, with extensive interaction among all levels of the corporate hierarchy and a strong prevailing team spirit. Prior to the move, the company carefully deliberated the decision to relocate. The old offices had sentimental value for many employees; this is where many had begun their careers. However, the new office building was easier to access via public transportation, allowed for increased company growth, and was more reflective of the company's success, with a striking exterior appearance and luxurious interior appointments.

Once the relocation was complete, however, many company members soon regretted it. The move carried with it some unintended and unfortunate consequences. The configuration of the new space was different: in the former building, executives' offices and cubicles were interspersed among staff cubicles, whereas in the new space all executives' offices were located on two separate floors. Initially, the concept behind establishing executive floors centered on easier communications among the directors; additionally, the company wanted to reward executives' performance by providing them with especially posh suites. However, some of the effects of the new configuration on the corporate culture were utterly unforeseen and, in the long run, damaging. The new configuration sharply reduced the casual interaction between executives and employees that had existed before, when many of the managers followed the strategy descriptively known as MBWA (literally, management by walking around). Employees and executives interfaced daily at the old building, meeting routinely at the coffee machines, in the hallways, and in the cubicles and offices; now, interaction was reduced to business-only discussions at formal meetings. Where before employees welcomed executives' casual visits and drop-ins in their cubicles, inviting the friendly and open banter, they now felt anxious and vaguely frightened during formal business meetings.

After some time in the new space, employees felt disconnected from company leadership, even disenfranchised from the company as a whole. There was now far less of a team atmosphere and more of a "them versus us" philosophy. Interestingly, executives too felt out of the loop with the employees, sensing a formidable new barrier between themselves and the staff. The team spirit that had once prevailed was sharply and unexpectedly curtailed by the office move; an important part of the company culture had been inadvertently and irrevocably destroyed.

new office, and about the processes and flow lines by which it will be accomplished. This is very similar to the flow charting commonly done by industrial engineers, especially when laying out material handling activities (Kroemer, 1997).

Theoretically, as pointed out earlier, there are two major issues to be considered in office space design: the work to be done and the personal needs and preferences of the employees doing the work. In reality, of course, other factors come into play, like budgets, location, timing, and even the status of people within the hierarchy of an organization. Many of us have worked in companies in which the square footage of individual offices corresponded in direct proportion to their inhabitants' position within the organization rather than to the actual tasks they performed. Yet, conceptually, what should be instrumental in designing office layouts are the tasks that will be done and the technical, physical, and psychological needs of the people who do them. As expressed by the chairman of a media company: "everyone knows who's important here, [we] don't need offices to establish that." (Iovine, 2000)

The process of office design involves a sequence of activities described below. What is critical here is that we must thoroughly analyze organizations before beginning office planning and design, taking a close look at the fundamental user needs. Office design is also very company-specific and office designs should be planned and created for each organization individually. At the core of an office's design is the analysis of processes, meaning what really goes on in an office and how work flow is conducted; accordingly, the resulting office design should reflect the actual communication needs rather than the boxes and lines of an often quasi-fiction organizational chart.

The process of office design consists of several important steps, to be taken one after the other.

1. Step 1 is the examination of what the people truly need who will inhabit the space. What tasks will they perform? What machinery and equipment will they use? Do they have specific preferences and work styles? How much do people need or want to interact? Is there a strict division of tasks? Is there a formal and strictly maintained hierarchy? What are the communication patterns among managers and personnel? We should also consider the company's management philosophy so that the appearance of the office can be matched to the company personality (see Chapter 2).
2. With this knowledge, the team of managers and users, architects, ergonomists, designers, and engineers can formulate specific statements of functional requirements. These statements guide the actual design.
3. The next step is to identify a range of design options so that practical solutions can be chosen; this reflects the need to compromise in the

real world since the conceptual ideal is not always possible. Constraints include time limitations, budgets, and political considerations.

4. Then, these candidate solutions are evaluated. For this, a formal process is best, involving careful definition and selection of criteria by which the candidate designs will be judged. The winning design is the one that garners the overall highest score.

5. At this point, the winning design is selected and implemented. Employees should be thoroughly trained here on any new office equipment or processes that the revamped design necessitates.

Note that these steps of office design apply, in principle, not only to the design of a new office but also to updating or remodeling an existing office. Note also that the definition and selection of criteria for judging the candidate designs eliminate the irrational decision making by the muddled "I like this better". The process outlined above is logical and transparent and forces all decision makers to follow the same rules.

Employees should be involved in *all* phases of the process so that their needs are truly met and they do not feel manipulated or ignored. If employees know that their inputs are taken seriously, then they will participate and cooperate.

Amenities should also be considered in office design. Nice features in the office space help attract new and retain current good employees, so a company involved in office build-out should carefully consider what kind of design amenities and perks it wishes to offer. Earlier, such options as well-equipped fitness centers with generous locker rooms and luxurious employee lounges with juice and coffee bars were mentioned; if such features are to be included, they should be considered when the initial layout is being planned because such options require special space configurations and hook-ups.

It is also critical that offices be flexible. In fact, many architects and expert designers treat office planning and layout as a continuing dynamic process. Consider the life cycle of a company. Most organizations wish to grow. Accordingly, they will need to plan for expanding office personnel and the requisite increasing demands for office space. One approach for keeping office designs flexible is to treat some parts of the office as more fixed than others and to keep the variable portions as easy to move or re-configure as possible. Kleeman (1991) referred to office landscape components as "shell" and "scenery". The building is the shell, which is difficult and costly and time-consuming to change. The scenery consists of the movable and easily changeable elements of interior design like partitions, furniture, and plants. Scenery can be replaced or revamped at relatively short intervals, and fairly inexpensively, reflecting the cycles of changing use in an office building.

Sample procedure to evaluate several possible office designs

Let us assume that several candidate designs for a new office have been submitted, designated here for short as Office 1, Office 2, Office 3, and Office 4.

These candidate designs are to be evaluated by a number of criteria: construction cost (ConstCost), ability to expand when needed (Expand), running costs (RunCost), appeal to employees and clients (Appeal), expected efficacy and effectiveness of use (EfficEffec), time that is expected to pass until the office can be occupied (Availability). Of course, there may be other criteria as well, but the following example will indicate how to insert them into the scheme.

The scoring, with1 worst and 10 best, is done by a panel of experts that includes managers and employees. In our case, the scores given by raters were as shown in Table 3.1. With all criteria of equal importance, according to the raw scores, Office Design #2 would be the chosen solution with 38 points, closely followed by Office # 4 with just one point less.

However, it is common that some of the criteria have higher importance than others. Accordingly, with weights (1–10, with 10 highest) applied to the criteria, the scores are as shown in Table 3.2. With weighted criteria, Office Design #4 is the preferred solution, with design #1 not far behind.

Table 3.1

	Office 1	Office 2	Office 3	Office 4
ConstCost	3	1	5	10
Expand	4	10	6	1
RunCost	9	10	1	5
Appeal	7	2	1	10
EfficEffec	8	10	5	1
Availability	1	5	8	10
Total	32	38	26	37

Table 3.2

	Office 1	Office 2	Office 3	Office 4
ConstCost (×8)	24	8	40	80
Expand (×3)	12	30	18	3
RunCost (×10)	90	100	10	50
Appeal (×7)	49	14	7	70
EfficEffec (×5)	40	50	25	5
Availability (1)	1	5	8	10
Total	216	207	108	218

Similarly, electric power and communication means must be flexible and moveable as well to accommodate growing staff size. This is made much easier by cordless phones, wireless computers, and by changes in attitudes of employees and managers.

New trends in office design

Some of the trends we see influencing office design today are centered around efforts to create a more home-like atmosphere and attempts to establish as much convenience as possible. Advances in technology are also profoundly affecting office designs today. These trends are interrelated.

Technology once tied us to one place – our workspaces needed to include wiring for electricity, data, and phone. Criss-crossing wires snaking over expanses of floor space to connect to distant outlets are hazardous and unacceptable in most situations (see Chapter 6). The former need for multiple and convenient connections constrained office design in the past. However, this is changing now with the advent and spread of wireless communications. "Our boss could just give us a handheld or laptop computer and a cell phone and tell us to work anywhere we wish." Even today, in the early spring of the year 2001, this is already reality for many of us – at least temporarily - as we work out of the office, while commuting, traveling by plane or train or car, waiting in the dentist's office, or in our home office (see below) and we may show up in the company office only once in a while. This has a profound impact on office design.

Thanks in part to new technologies, at present there is an active movement to design offices that free workers from their cubicles. This allows for personal direction and individuality at the office and also enables companies to create environments that are more like home. Whether it involves providing child care facilities or adding conveniences like on-site wellness centers, masseurs, and dry cleaners, employers will have to continue to make it easier for people to be at their jobs. Not only are floor plans reflecting more innovative space use, but companies are using natural materials and colors to achieve a more biological, natural feel. The bee hive and the paperwork factory are out, free enterprise and telecommuting are in. All of this can contribute to job satisfaction – which (as discussed in Chapter 2) may encourage people to work better and is likely to enhance their performance and productivity.

The home office

Following one's individual interests and preferences in the layout of the workspace is easiest in the home office – the personal workstation away from the company building. As mentioned in Chaper 1, we are free to arrange our own workspace just as we like: spartan or luxurious, strictly ad hoc or using concepts such as feng shui.

Many of the tasks comprising office work are now done outside of the traditional office in the company building. We now often work on a laptop or handheld computer or with a wireless phone wherever we happen to be, frequently at times other than the old nine-to-five office hours, and increasingly in the private office at home. Telecommuting and working at home are more and more common, yet the term home office still has the connotation of a corner in the den, a spot in the kitchen or, at best, a spare room. That inferred meaning also implies the use of furniture that is not specifically designed for office work, like a folding chair at a card table or an easy chair in that corner of the den. However, as the occasional work changes into schedules of workdays with long hours of effort, the room or space housing the home office should follow the ergonomic design recommendations we mention throughout this book. Moreover, the home office should adhere to the office space guidelines we examined in this chapter.

Depending on one's specific work at home, the workspace must accommodate various tasks: that may be drafting, generation of samples, storage of materials, filing of documents, shelving of literature. One may run a complete business from the private office, with all its organization and administration, bookkeeping, faxing, e-mailing and telephoning. Or one may just need room for a computer workstation. In any case, the home office should be carefully selected, suitably cooled or heated, properly insulated and protected from noise and interference, and well lit. Of course, we should closely examine the individual needs and preferences of the worker before we design the home office space. Chapters 4 and 5 give information about computers and furniture that applies to the home office, and Chapters 7, 8, and 9 discuss aspects of lighting, sound and climate that are also of importance.

ERGONOMIC DESIGN RECOMMENDATIONS

When designing office space layout, take a systematic approach:
- Identify and understand the needs, capabilities, and limitations of the people who will occupy the space.
- Carefully consider the work itself. It may include confidential or highly-focused tasks that require privacy, or it may require face-to-face contact with other people.
- Take into account real world constraints like size limitations of the space, budget constraints, hierarchies that must be observed, and time schedules.

Specific environmental conditions that you should address in designing office space include:
- Noise – consider the acoustical requirements in offices; we should be able to perform normal office work with minimal disruption or distraction from standard office sounds like another worker's voice. Sound absorbing screens, carpets, draperies, outer shells of filing cabinets can all help to reduce noise.
- Lighting – the office environment should allow us to see clearly and without glare; task lighting is often the most suitable; we should also consider natural light when feasible.
- Climate control (especially of temperature and humidity) influences our wellbeing and performance.

Other considerations include:
- Since appealing office designs may help attract employees, plan in advance for amenities the company may wish to offer.
- Consider also what the office's design expresses about the corporate character or personality of an organization.

Follow these steps to establish the most effective and efficient design for your office:
- Analyze the needs of the workers who will inhabit the office, including the tasks they will perform, the machinery and equipment they will use, and their preferences and work styles; this includes looking at the amount of interaction and the level of bureaucracy in the business as well as examining the company's management philosophy.
- Formulate specific statements of the functional requirements you determined in the first steps; these statements will guide the actual design.
- Identify a range of design options so that practical solutions can be chosen.
- Evaluate these options by a set of defined criteria; see Tables 3.1 and 3.2.
- Select the final design.
- Implement the design, providing plenty of training and support for employees as you go.

Keep in mind these important guidelines as you plan the office layout:
- Get employees involved in *all* steps of the process so that their needs are truly met and they do not feel manipulated or ignored; you can do this by soliciting their input, asking for their opinions, and giving them any training they need on any new space or equipment. Ideally, users of the new space would actually be co-designers.
- Make the office space as flexible as is feasible; the organization will likely change in the future and require plenty of variability in space

and its layout; one way to accomplish this is through modular panel systems and movable dividers.

- Remember that flexibility extends to power and communications sources, which should also be easily movable, unless the office and its employees are wireless.

References

Arnold, D. (1999). The Evolution of Modern Office Buildings and Air Conditioning. *ASHRAE Journal 41* (6), 40–54.

Hedge, A. (2000). Where Are We in Understanding the Effects of Where We Are? *Ergonomics 43*, 1019–1029.

Iovine, J. V. (2000). Playful Ideas Infuse Modern Interior Design. *Commercial Real Estate Supplement of the Chicago Tribune, 29 March*, 15.

Kleeman, W. B. (1991). *The Challenge of Interior Design*. New York: Van Nostrand Reinhold.

Kroemer, K. H. E. (1997). *Ergonomic Design of Material Handling Systems*. Boca Raton, FL: CRC Press/Lewis.

Raymond, S. and Cunliffe, R. (2000). *Tomorrow's Office*. London: Taylor & Francis.

4   Chairs and other furniture

Overview

The last 300 years

Nearly 300 years ago, in 1713, Ramazzini stated that workers who sat still, stooped, looking down at their work (such as tailors) often became round-shouldered and suffered from numbness in their legs, lameness, and sciatica. Ramazzini generalized that "all sedentary workers suffer from lumbago," and he advised workers not to sit still but to move the body, and "to take physical exercise, at any rate on holidays" (Wright, 1993, pp. 180–185).

In the offices of the late 1800s and the early 1900s, when the clerks were male, it was common to stand while working. Then the concept changed and sitting in the office became customary. Yet, low back pain and musculoskeletal irritation, often together with eyestrain, are common complaints of persons who operate computers or do other tasks while sitting in the office. Liberty Mutual reported in 1999 that hand, wrist and shoulder disorders were a fast-growing source of disability in the American workplace, stemming in large part from the dramatic increase in office technologies in the latter part of the twentieth century. "As computers have become a staple in the workplace, work-related musculoskeletal irritation has increased." (Liberty Mutual, 1999, p. 14.) This is a serious, disappointing, and utterly avoidable development that runs counter to all ergonomic knowledge.

Our body is built to move about, not to hold still

It is tiresome to maintain any body position unchanged over extended periods of time. This includes sitting still for hours on end. We would like to move about, but in the computer office we are double-tied to our workstation: by our hands, which must operate the keys, and by our eyes, which must view screen, text, and keys. These two ties keep our hands and eyes and, through them, our whole body fixed at the workstation. We try to ease the effort of maintaining the rigid posture by sitting on a chair as comfortably as practicable, shifting and slouching as needed, but our body keeps telling us to "get up and walk and move around" for a while.

Sitting in the office – but how?

The modern office has little resemblance to the rooms a century ago in which clerks labored. Then, the clerks were men who stood at their desks, using ink to write letters and hand-printing entries in ledgers. By the middle of the twentieth century, clerks had changed from standing at work to sitting, and most office employees in clerical roles were females. The idea of "erect sitting is healthy sitting" had prevailed over standing upright, and office furniture was designed for this body position.

About 120 years ago, body posture had become of great concern to physiologists and orthopedists. In their opinion, the upright (straight, erect) standing posture was balanced and healthy while curved and bent backs were unhealthy and therefore had to be avoided, especially in youngsters. Consequently, "straight back and neck, with the head erect" became the recommended posture for sitting and, logically, seats were designed to bring about such upright body position.

For about a century, office furniture was purposely made for people to maintain their bodies in that erect posture. But that applied only to clerical employees: managers habitually enjoyed an ample armchair with high back and comfortable upholstery while the secretary sat on a small hard chair with a miserable little board for a back rest.

Feeling comfortable

The simplistic concept that sitting upright, with thighs horizontal and lower legs vertical, meant healthy sitting lasted, surprisingly, for about 100 years. Now it is obvious that people in modern offices sit any way they like – not only without bad health consequences, but apparently because freely choosing and changing their posture makes them feel comfortable.

Sitting, as opposed to standing, is suitable when only a small work-space must be covered with the hands; this is typical for much of today's office work. Sitting keeps the upper body stable, which is helpful when finely controlled activities must be performed. Sitting supports the body at its mid-section and requires less muscular effort than standing, especially when maintained over long periods of time. But the seat must be supportive to the body, feel comfortable in combination with the other office furniture and equipment, and be suitable for the work tasks.

New work duties, the widespread use of computers, and changing attitudes give reasons to rethink traditional design recommendations for office furniture. The furniture should accommodate a wide range of body sizes, varying body postures and diverse activities; it should enhance task performance, facilitate vision and allow interaction with co-workers; it should be appealing and help make people feel well in their work environment.

Ergonomic recommendations for proper design of workstations and furniture, especially of the chair, are at hand to make work easy and efficient.

> You may skip the following part and go directly to the Ergonomic Design Recommendations at the end of this chapter – or you can get detailed background information by reading the following text.

What, exactly, are "good" body movements and "healthy" postures?

It is an everyday experience that maintaining any body position becomes unpleasant after a while: even while we rest in bed, or relax in our most cozy easy chair, we must move our body and re-position ourselves after some time to remain comfortable. The need to change the body's carriage is urgent after we have been forced to sit still, such as in a crowded airplane, or worse, after standing still, such as at attention in a military formation. Our body is built to move about, not stand or sit still.

Consequently, we should design our workstations for movement, but unfortunately, design for an imaginary good maintained posture has been pursued – with little success – for about a century now.

Theories of healthy postures

In the nineteenth century, body posture was of great concern to physicians, physiologists, and orthopedists. In 1889, the German orthopedist Staffel reported that farmers and laborers often had back curvatures that diverted from the norm: their spinal columns were either too flat, or overly bent – lordotic (backward), kyphotic (forward), and scoliotic (sideways). Staffel and his contemporary orthopedic experts found these back postures unhealthy and concluded that they had to be avoided, especially in children.

After observing men standing easy and unconstrained, head high and looking straight forward, Staffel classified standing postures according to their spinal curvatures. He characterized a normal posture by a straight spine in the frontal (or rear) view. Seen from the side, a plumb line from the top of the cranium passes through the cervical vertebrae, the shoulder joint, the lumbar vertebrae; below, the plumb line falls just behind the center of the hip joint and then down the leg where it stays slightly in front of the centers of the knee and ankle joints. This posture, upright and straight, shown in Figure 4.1, appeared balanced and healthy to Staffel.

For nearly a century, physicians, orthopedists, physical therapists, parents, teachers and military officers advocated the nineteenth century normal posture of erect standing, as recommended by Staffel and his contemporaries. Even today, that upright standing posture is commonly considered good and proper. But it is certainly impractical for working since we do move about – and should do so – instead of standing (or sitting) stiff and still.

In recent years, the term "neutral posture" has become popular, suggesting a healthy or desirable or middle position of body members. Unfortunately, it is often not clear what "neutral" means; is that the middle of the total motion range in a joint? This would make some sense for the wrist, indicating the hand is straight, in line with the forearm. But there is no obvious significance to the middle joint position in the elbow or knee, shoulder or hip, or the spinal column. Does the term neutral suggest that all tissue tensions around a joint be balanced, so that the position is stable? Does the term imply a minimal sum of tissue tensions (torques) around a body joint? Or does this apply to tensions

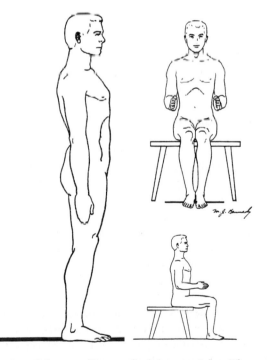

Figure 4.1 Standing straight standing and sitting upright. These static body postures are useful for classic body measurements, but they are not working postures.

about several joints, or all body joints? Does neutral imply minimal joint discomfort (Genaidy and Karwowski, 1993)? Does it refer to a relaxed posture? Or a posture instinctively assumed for a task, to generate high body strength, or to avoid fatigue?

Sitting upright

A few years before his treatise on standing postures, in1884, Staffel had published his theories about "hygienic" sitting postures. He recommended holding trunk, neck and head erect, with only slight bends in the spinal column in the side view. His recommendation for the desired back posture when sitting was similar to what he advocated for standing a few years later.

Staffel and his contemporary colleagues were particularly concerned about the postural health of children and therefore, as they all agreed, school seats and desks should be designed, and the children exhorted, to maintain that erect posture of back, neck and head. Starting in the late 1880s, a great number of hygienic, healthy designs for school furniture were proposed: seats, desks, and seat–desk combinations laid out to promote the upright posture (Bonne, 1969; Merrill, 1995; Zacharkow, 1988). Figure 4.2 shows Schindler's 1890 design of school furniture.

The same sitting posture recommended for pupils was also advised for adults: head, neck and trunk upright, with thighs horizontal and lower legs vertical. For much of the twentieth century, office furniture was endorsed that attempted to make people sit in that erect posture; even the 1988 ANSI/HFS Standard 100 employed it to prescribe office furniture (see Figure 4.3). It is curious, however, that the sitting posture with 90° angles in the joints of ankles, knees, and hips was mostly expected from lower-level office personnel. The boss and his managers habitually enjoyed an ample armchair, upholstered, with a high backrest made to recline for comfortable relaxation (Bradford and Byrne, 1978; Tenner, 1997). Compare your experience of the sustained stilted upright position to the feeling of sitting in an easy chair that rocks.

The simplistic concept that sitting upright, with thighs horizontal and lower legs vertical, means sitting healthily endured for a surprisingly long time. Even today, that upright posture with slight lordoses (forward bends) in the lumbar and cervical spine areas and a light kyphosis (backward bend) in the thoracic spine is typically considered healthy, balanced or neutral (Merrill, 1995). Obviously, this posture can be quite appropriate for a while, but it is erroneous to use it as the overriding guiding principle for the design of the chair, and of other workstation furniture. To simply design for this postural idol completely disregards that it is healthy to change among various postures while sitting, rather than to stay in any one position.

It is very unfortunate that the conventional design of the computer

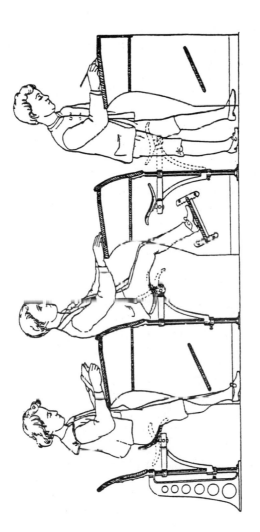

Figure 4.2 Schindler's 1890 design of school furniture. It contains about everything that can be thought of for body support in semi-sitting, sitting or standing at work.

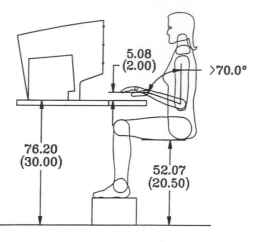

Figure 4.3   Title page of the 1988 ANSI/HFS Standard 100. Compare to Figure 4.4.

workstation also implies one maintained posture; the currently prevailing idea seems to be that the hands must be on the keyboard, the eyes fixated on the computer screen. The ensuing positions of hands and arms, eyes and head determine the position of the upper body, not allowing much variation in posture (Veiersted, 1998).

Imagine the freedom to place the body if a keyboard were not needed, or if at least it were not stationary, and if the display were large and moveable: such design features are technically feasible. There is more discussion of these aspects in Chapters 5 and 7 (and in many other publications, e.g., Kroemer et al., 2001) but even now it is intuitively

Figure 4.4   Grandjean's 1987 characterization of the expected upright posture of the computer operator: wishful thinking.

clear that no one and only healthy posture exists for which the work-station should be designed; even in his 1987 book, Grandjean (p. 149) called this idea "wishful thinking" (see Figure 4.4).

Assessment of suitable postures

As mentioned above, for about a century it was commonly believed that standing or sitting with a straight back is physiologically desirable and socially proper. Of course, there is nothing wrong with upright sitting or standing at one's own will, but it is erroneous to require that an erect back be maintained for long periods of time, such as during work while sitting in the office. Obviously, the human body is made for change, to move about. Sitting (or standing) still for extended periods of time is uncomfortable; it leads to compression of tissue, reduction in metabolism, deficiency in blood circulation, and accumulation of extracellular fluid in the lower legs.

In his book, *Sitting on the Job*, Donkin (1987, p. 12) asked: "If slumping in my chair is so bad for me, why does it feel better at times?" The answer lies in the fact that slumping and other body movements are the instinctive attempts to take strain and tension away from muscles that are working to maintain prolonged postures (Aaras et al., 1997; Halpern and Davis, 1963; Michel and Helander, 1994; Williams et al., 1991).

Techniques to assess the effects of changes in posture

The effects of postural changes can be measured and evaluated by observing variations in dependent variables (Kroemer and Grandjean, 1997; Kroemer et al., 2001). Researchers from several disciplines can measure these effects using their specific techniques:

- Physiology: oxygen consumption, heart rate, blood pressure, electromyograms, fluid collection in the lower extremities;
- Medicine: acute or chronic disorders including cumulative trauma injuries;
- Anatomy and biomechanics: X-rays, CAT scans, changes in stature, disc and intra-abdominal pressure, model calculations;
- Engineering: observations and recordings of posture; forces or pressures at seat, backrest, or floor; amplitudes and frequencies of body displacements; productivity;
- Psychophysics: structured or unstructured interviews, subjective ratings by either the experimental subject or the experimenter.

Some of these techniques are well established and easy to use, but many are not. Most require a laboratory setting, some are suitable for field studies. For practically all outcomes, however, the threshold values that separate suitable from unsuitable conditions are unknown or variable. Thus, the interpretation of the results obtained by most of the listed techniques is difficult, to say the least.

Subjective assessments, questionnaires

Subjective judgments presumably encompass all the phenomena addressed in the various measurements listed above, and they can be reliably scaled and interpreted (Booth-Jones et al., 1998).

Based on initial work by Shackel et al. (1969) and Corlett and Bishop (1976), a large number of questionnaires have been developed for assessing people's feelings about unsuitable or painful working conditions (for a recent overview, see Kroemer et al., 2001). An often used, well-standardized inquiry tool is the Nordic Questionnaire by Kuorinka et al. (1987), modified by Dickinson et al. (1992). It consists of two parts, one asking for general information, the other specifically focusing on low back, neck, and shoulder regions. It uses a sketch of the human body, divided into nine regions (see Figure 4.5). The interviewee indicates any symptoms that may exist in these areas. If needed, more detailed body sketches can be used (van der Grinten and Smitt, 1992).

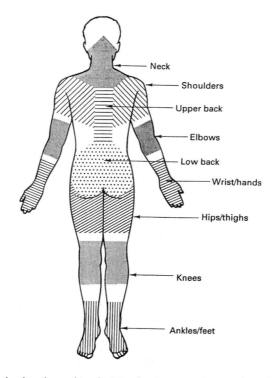

Figure 4.5   Body sketch used in the Nordic Questionnaire. The subject looks at the drawing while identifying painful spots.

The Nordic Questionnaire probes deeply into the nature of complaints about pains and discomfort, their duration, and their prevalence.

Comfort versus discomfort

Comfort (as related to sitting) has long been defined, simply and conveniently, as the absence of discomfort. However, the underlying concept is false: Helander and Zhang (1997) showed that, in reality, these two aspects are not the opposite extremes on one single judgment scale. Instead, there are two scales, one for the agreeable feelings of comfort and the other for such unpleasant experiences as not being at ease, fatiguing, straining, smarting, hurting – all terms that indicate some degree of discomfort. These two scales partly overlap but are not parallel.

To avoid the false concept of one scale that has comfort and discomfort as polar opposites, in this text we use the term *annoyance* (instead of discomfort) as the descriptive label for the scale containing the unpleasant statements. The other scale, containing the agreeable statements, will be labeled with the term *comfort*, as has been the convention.

The annoyance scale

Feelings of annoyance (formerly called discomfort) are expressed by such words as stiff, strained, cramped, tingling, numbness; not supported, fatiguing, restless, soreness, hurting, and pain. Some of these attributes can be explained in terms of circulatory, metabolic, or mechanical events in the body; others go beyond such physiological and biomechanical phenomena.

Users can rather easily describe design features that result in feelings of annoyance such as chairs in wrong sizes, too high or too low, with hard surfaces or sharp edges; but avoiding these mistakes does not, per se, make a chair comfortable.

The comfort scale

Feelings of comfort when sitting are associated with such descriptive words as warm, soft, plush, spacious, supported, safe, pleased, relaxed and restful. However, exactly what feels comfortable depends very much on the individual and his or her habits, on the environment and task at hand, and on the passage of time.

Esthetics plays a role: if we like the appearance, the color, and the ambience, we are inclined to feel comfortable. Appealing upholstery, for example, can strongly contribute to the feeling of comfort especially when it is neither too soft nor too stiff but distributes body pressure along the contact area, and if it breathes by letting heat and humidity escape as it supports the body.

Ranking chairs by annoyance or comfort

Helander and Zhang used six specific statements about chair annoyance or comfort (each with nine steps from 'not at all' to 'extremely') followed by one general statement.

The statements for annoyance were as follows:

1. I have sore muscles
2. I have heavy legs
3. I feel uneven pressure
4. I feel stiff
5. I feel restless
6. I feel tired
7. *I feel annoyed*

The following statements characterized comfort:

1. I feel relaxed
2. I feel refreshed
3. The chair feels soft
4. The chair is spacious
5. The chair looks nice
6. I like the chair
7. *I feel comfortable*

Helander and Zhang's subjects said it was easy to rank chairs in terms of *overall comfort* or *annoyance* (answer no. 7 in the lists) after having responded to the preceding more detailed descriptors nos. 1–6.

The researchers concluded that it is apparently more difficult to rank chairs, unless truly unsuitable, by attributes of annoyance (as opposed to comfort) because the body is surprisingly adaptive except when the sitter has a bad back. In contrast, comfort descriptors proved to be sensitive and discriminating for ranking chairs in terms of preference.

It is also of interest to note that preference rankings of chairs could be established early during the sitting trials; they did not change much with sitting duration. Still, it is not clear whether a few minutes of sitting on chairs are sufficient to assess them, or whether it takes longer trial periods.

Free flowing motion

Grandjean et al. (1984) found that persons sitting in offices did not sit upright but leaned backward even if their chairs apparently were not designed for such a posture. Bendix et al. (1996) reported that persons, while reading, often assumed a kyphotic lumbar curve even when sitting on a chair with lumbar pad that should have produced a lordosis. Apparently, people sit any way they want, regardless of how experts think they should sit!

Allowing persons to freely select their posture has led in two instances to surprisingly similar results. In 1962, Lehmann showed the contours of five persons resting under water where the water fully supports the body. Sixteen years later, NASA astronauts were observed when they relaxed in space. The similarity between the postures under water and in space is remarkable, as shown in Figure 4.6. One might assume that, in both cases, the sum of all tissue torques around body joints has been nulled. Apparently not by accident, the shape of so-called easy chairs is quite similar to the back contours shown in both figures. Some executive office workstations allowed the boss, when he or she leaned back, to assume similar body shapes.

Dynamics is a label than can be applied to current design of office chairs, as opposed to the statics of maintained posture. People do move around. Design should encourage and support free flowing motions, as shown in Figure 4.7, with opportunities for temporary postures at the whim of the person.

The *free-flow motion* or *floating support* design idea has these basic tenets:

- Allow the user to freely move in and with the chair and to halt at will in a variety of sitting postures, each of which is supported by the chair; and to get up and move about.
- Make it easy for the user to adjust the chair and other furniture, especially keyboard and display, to the changing motions and postures.
- Design for a variety of user sizes and user preferences.
- Consider that new technologies develop quickly and should be usable at the workstation. For example, radically new keyboards and input devices, including voice recognition, may be available soon; display technologies are undergoing rapid changes; wireless energy transmission may no longer limit the placement of display and input devices.

Figure 4.6   Relaxed postures under water and in low gravity.

Figure 4.7   Free flowing motions: people differ in size and preferences, and everybody changes, moves, gets up – does anybody ever sit still for long
periods of time?

Design of the office workstation

Successful ergonomic design of the workstation in the office depends on
proper consideration of several interrelated aspects, shown in Figure 4.8.
Work tasks, work movements, and work activities all interact. They
affect, and are influenced by, the workstation conditions, including furniture and other equipment, and the environment. All of these must fit the
person to achieve individual wellbeing and foster high work output. Of
course, job content and demands, control over one's job and many other
social and organizational factors also influence feelings, attitudes and
performance (see Chapter 2).

When designing the layout of a work task and workstation, it helps to
consider three main links between a person and the task.

- The first link is the visual interface: one must look at the keyboard,
the computer screen or the printed output, and source documents.
- The second link is manipulation: the hands operate keys, a mouse or
other input devices; they manipulate pen, paper, and telephone.
Occasionally, the feet operate controls; starting and stopping a dicta-

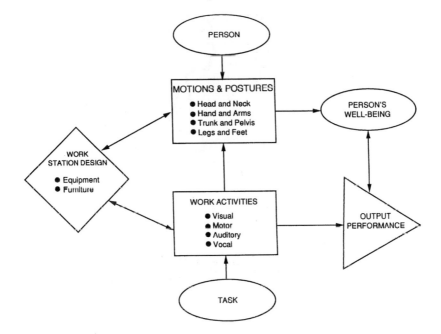

Figure 4.8 Person, task, workstation and performance.

tion machine is an example. The intensities of the visual and motor requirements depend on the specific job.

- The third link is body support: the seat pan supports the body at the undersides of the thighs and buttocks, and the backrest supports the back. Armrests or a wrist rest may serve as support links.

Designing for vision

The location of the visual targets greatly affects the body position of the computer operator (Ankrum and Nemeth, 1995; Bauer and Wittig, 1998; Dowell et al., 1997; Hagberg and Rempel, 1997; Hamilton, 1996; Hill and Kroemer, 1986; Kroemer, 1985; Turville et al., 1998; Villanueva et al., 1996). Many studies have demonstrated what is intuitively apparent: objects upon which we focus our eyesight should be located directly in front, at a convenient distance and height from the eyes. If one is forced to turn the head to the side or tilt it up to view the computer screen or the document, eye strain is commonly experienced, often together with pain in neck, shoulders, and back. Yet, it is surprising to observe that two basic mistakes are often made in many workstations: the monitor is set up much too far away and too high, as in Figure 4.9; and source documents are often laid flat to the side. In either case, the

Figure 4.9   The monitor is set up too far and too high, so the operator arches back and neck trying to get the image into focus.

operator must crane the neck. See Chapter 7 for how to select the proper line of sight and viewing distance.

Monitor support

Whenever possible, use a separate support structure (table or stand) for the computer monitor so that the display can be adjusted in height independently from work surface and keyboard. Easy adjustment is facilitated if the support is spring-loaded and can be moved up or down by a hand crank or, even better, by an electric motor.

The all too common practice of putting the monitor on the CPU box, and possibly also on a stem for angle adjustment, lifts the screen much too high for most users who, as a consequence, tilt their head back and then often suffer from neck and back problems. Instead, the monitor should be located low behind the keyboard so that one looks down at it.

As a rule, the screen or source document should be about half a meter from the eyes, the proper viewing distance for the operator. This is the reading distance for which corrective eye lenses are usually ground. A convenient yardstick is to place the screen and source document at arm's length, or slightly less.

Figure 4.10   A document placed far to the side causes a twisted body posture and lateral eye, head and neck movements (modified from a sketch provided courtesy of Herman Miller, Inc.).

Document holder

If you often read from a source document, use a document holder to hold the document close and parallel to the monitor screen, about perpendicular to the line of sight. A document placed far to one side causes a twisted body posture and lateral eye, head and neck movements (see Figure 4.10)

Corrective eye lenses

When you experience eye strain even though the workplace seems in good optical order, it is time to visit an ophthalmologist or optometrist to have your vision checked. That specialist may determine, to one's dismay, that an eye correction is needed, often in form of corrective lenses. This is rather likely to be the case when the operator has reached middle age, when natural lenses regularly lose their ability to focus on close visual targets (see Chapter 7), which makes it difficult to discern characters on the computer screen or in a source document. It is fairly easy to resolve that problem by wearing artificial lenses, either as eyeglasses or contact lenses, which will take over the task of focusing on targets at reading distance. In bifocals or trifocals, the lenses for reading are ground into the lowest section, because reading is naturally done by looking downward to text, as discussed further in Chapter 7.

Having the reading section in the lowest segment of the glasses makes it easy to read a document, or text on the screen, when they are placed low into your field of vision. But if these targets are put up too high, then you must severely tilt your head up to see them (see Figure 4.11). The bent-back neck often leads to headaches.

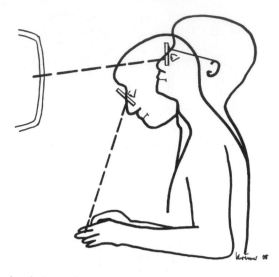

Figure 4.11   Bifocal glasses have their reading lenses at the bottom. Looking up to the display causes severe backward tilt of the head, extreme neck kyphosis – and probably a piercing headache.

Proper lighting

The computer screen is self-lit, assuming it was manufactured according to the newest standards such as ANSI/HFES 100. Overall, the computer office needs to be illuminated at about 200–500 lx. Paper documents may be difficult to read at this fairly low level, so one might want to shine a special task light on them; make sure, however, that this does not create glare. For offices without computers, the proper illumination range is from 500 to 1000 lx, even more if there are many dark (light-absorbing) surfaces in the room. Chapter 7 discusses lighting in detail.

Designing for manipulation

In addition to the eyes, our hands are usually very busy doing various office tasks: grasping and moving papers, taking notes, punching phone numbers, and using various computer input devices. If our hands are engaged in many different activities, the varied manipulation is likely to keep our arms and the upper body moving around in our workspace. Motion is desirable, in contrast to maintaining a fixed posture, such as when tapping on the keyboard over extended time or during lengthy "mousing".

When sitting, we do have a large manipulation area available, especially if we move the upper body and, of course, we can cover an even larger area when standing up. However, for finely controlled hand movements, we prefer a space near chest and belly, about 10–40 cm in front of the body (see Figure 4.12). Work done here can also be seen acutely because objects are at a suitable distance from the eyes, and so low as to fit the natural downward direction of gaze (see Chapter 7).

Chapter 5 contains an extensive discussion of keyboards and other computer input gadgets, but remember that these devices should all be placed directly in front of our body, at about elbow height when the shoulders are relaxed and the upper arms hang to our sides. A keyboard

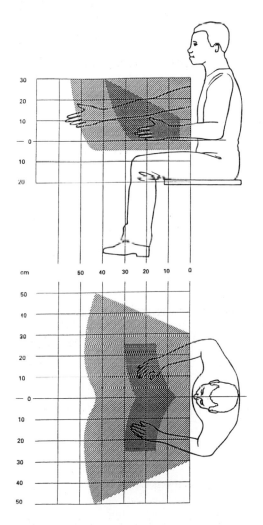

Figure 4.12 Areas for gross movements of the hands of a sitting person and, cross-hatched, for finely controlled manipulations such as hand writing or keyboarding.

set up higher makes us lift arms and shoulders, requiring unnecessary muscle tension (see Figure 4.13). Placing keyboard and mouse at different levels, for example, or to the side, or too far away, will cause muscle tension in the back, neck, shoulders, and arms to get and hold and move the hands there. This often leads to irritation and pain, even occasionally to overuse disorders such as bursitis, tendinitis or cervico-brachial syndrome (see Chapter 6).

Resting one's wrist or arm on a hard surface or, worse, on a hard edge often occurs when a working surface is pushed up too high, above the operator's elbow height. This leads to sharp local pressure at the point of contact that can cause painful reactions, as Figure 4.14 shows. Examples are cubital tunnel syndrome when the cubital nerve is compressed by placing the elbow on a hard surface, or carpal tunnel syndrome when a sharp edge (of a keyboard housing, for example) presses into the hand's palm area near the wrist joint. These conditions can be avoided by choosing a low manipulation area, by softening surfaces that support arm and hand (rounding edges, padding surfaces, providing wrist supports) and by adopting proper working habits.

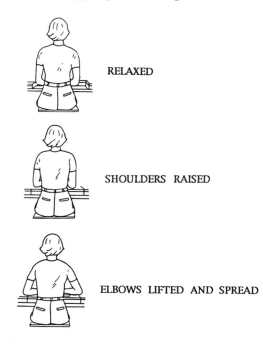

RELAXED

SHOULDERS RAISED

ELBOWS LIFTED AND SPREAD

Figure 4.13   Keyboard height affects body posture. With the keyboard just at the right height above the seat, shoulders and arms hang relaxed. A keyboard set up too high requires unnecessary muscle tension to lift the shoulders and arm and even to lift and spread the elbows. (This sketch is modified from Grandjean's illustration showing Hagberg's research results; see Grandjean, 1987, p. 104.)

Figure 4.14   High local pressure spots at sharp and hard surfaces. (Modified from a sketch provided courtesy of Herman Miller, Inc.)

Designing for motion

We have mentioned this several times now but will reiterate it again: our body is built to move about, not to hold still. It is uncomfortable and tiresome to maintain a body position without change over extended periods. We experience this while driving. The car driver must hold the head in much the same position in order to see road and instruments, keep hands and foot fixed on wheel and pedal. After a while, it becomes very difficult to stay in the same seated position and posture, even when a rather comfortable seat supports the nearly immobile posture. A similar situation arises often at computer workstations, where the chair is often much less suitable than our car seat. (One can buy certain car seats adapted for use in the office). But, unlike when driving, we can get up and move around at will in our office.

Since the primary aim of ergonomic workstation design is to facilitate body movement, the designer should consider the extreme body postures expected to occur and lay out the interim workspace for motion. In the computerized office this means to design for walking (standing) and for sitting; the implication for the seat is that it should be designed for relaxed and upright sitting, for leaning backward and forward, and for getting in and out.

Several chair designs have gone beyond the old concept of user adjust-ment. They incorporate interesting ideas, like having seat pan and back rest follow the motions of the sitting person and providing support throughout the range, with pan and backrest either moving indepen-dently or in a linked manner. Other designs start from the premise that the seat must not be a passive device but an active one: the chair as a whole, or its pan or backrest, can automatically change the configuration slightly over time, perhaps in response to certain sitting postures main-tained by the person. The change can be in angles, or in stiffness of the material. Seat and back cushions that pulsate were tried in the 1950s to alleviate the strain that military aircrews felt when they had to sit for hours on end to fly extended missions. Should we be reminded by an "intelligent" seat to move, or to get up, after we sat in a static position for some period of time?

In terms of the general office design, it would be desirable to provide several workspaces for every person. One workspace is the conventional sit-down desk and computer station. An optional stand-up station can be another workspace. Then there are separate areas for filing written materials, another spot for supplies, rooms to meet with colleagues and visitors, to have a cup of tea or coffee or soft drink and perhaps a nap (see also Chapter 3). Moving about can be encouraged, even designed into the office (see Figure 4.15) by having files stored in a place away from the desk, for example. However, having to twist and contort the body, like in Figure 4.16, because of bad office layout, or just because of self-made clutter, is not a healthy idea.

Figure 4.15 It is good to get up and stretch the body. (Modified from a sketch provided courtesy of Herman Miller, Inc.)

Figure 4.16   Awkward postures do not usually mean good exercise. (Modified from a sketch provided courtesy of Herman Miller, Inc.)

Designing the sit-down workstation

One of the first steps in designing the office workstation for seated persons is to establish the main clearance and external dimensions. The size of the furniture derives essentially from the body dimensions and work tasks of the people in the office. Main vertical anthropometric inputs to determine the height requirements are lower leg (popliteal and knee) heights, thigh thickness, and the heights of elbow, shoulder and eye. (For more details see, for instance, the publications by Dainoff, 1999; Kroemer et al., 1997, 2001; Pheasant, 1996.)

The common design procedure is to start from the floor and add the height of the chair, then the height of supports for input devices (such as the keyboard, mouse pad, etc.), finally the height of table and desk surfaces. To truly fit all the sizes and preferences of everyone in the office, all the furniture heights should be widely, and easily, adjustable.

Another furniture design strategy starts with the fixed height of the major work surface (the desk or table in traditional offices), with adjustable heights of the seat and of the computer support. This usually necessitates narrower height adjustment ranges for seat and equipment, but smaller people will need footrests. Still another design approach relies on the same seat height for all, which results in the smallest adjustment needs for desks, tables, and supports, yet all but the most long-legged individuals need footrests (Kroemer et al., 2001).

The depths and widths of the furniture must fit the horizontal body dimensions (especially popliteal and knee depths, hip breadth and reach capabilities) as well as work task and equipment space needs. ANSI/ HFES 100, in its newest edition, provides some help for selecting appropriate furniture dimensions.

The furniture at the computer workstation consists primarily of the seat, the support for the data entry device, the support for the display and a working surface. It is best, and most expensive, to have all of these independently adjustable. Recommended approximate height adjustment ranges for office furniture in Europe or North America are as follows:

- seat pan above the floor: 37–51 cm, better up to 58 cm
- support surface for keyboard, mouse, etc.: 53–70 cm
- surface of worktables: 53–70 cm
- desk surface: 53–72 cm
- support for the display: 53–90 cm

These ranges should make the office furniture fit practically everybody, tall or short. Of course, if the work place is used by one individual, such as a personal home office, then just that one user must be fitted and little or no adjustment may be necessary. When applying these recommended ranges for adjustment keep in mind that people vary greatly in their body sizes, within one population group, and certainly from continent to continent.

The office chair

As discussed earlier, proper sitting at work was long believed to mean an upright trunk, with the thighs (and forearms) in essence horizontal and the lower legs (and upper arms) vertical. This model with all major body joints at 0, 90 or 180° makes for a convenient but misguiding design template: the 0–90–180 posture is neither commonly employed nor subjectively preferred, and it is not even especially healthy. To suit the seated person, the designer of office furniture, especially of chairs, must consider a full range of motions and postures, as discussed earlier.

Seat pan   When one sits down on a hard flat surface, not using a backrest, the ischial tuberosities (the inferior protuberances of the pelvic bones) act as fulcra around which the pelvic girdle rotates under the weight of the upper body. Since the bones of the pelvic girdle are linked by connective tissue to the lower spine, rotation of the pelvis affects the posture of the lower spinal column, particularly in the

This text contains specific design recommendations, for example in chair dimensions, which have been derived to fit mostly North American and European users. Their body sizes are fairly well known but other populations with widely different body sizes would need their own ranges in furniture dimensions and adjustments; see the compilations of international anthropometry by Pheasant (1996) and Kroemer et al. (1997). Furthermore, working postures and habits differ widely among the various peoples on earth, as described, for example, by Kroemer et al. (2001) and Nag et al. (1986). Therefore, the recommendations given here most likely need to be modified to suit users in other parts of the globe.

lumbar region. If the pelvis rotation is rearward, the normal lordosis of the lumbar spine is flattened (see Figure 4.17).

Leg muscles (hamstrings, quadriceps, rectus femoris, sartorius, tensor fasciae latae, psoas major) run from the pelvis area across the hip and the knee joints to the lower legs. Therefore, the angles at hip and knee affect the location of the pelvis and hence the curvature of the lumbar spine. With a wide-open hip angle, a forward rotation of the pelvis on the ischial tuberosities is likely, accompanied by lumbar lordosis. (These actions on the lumbar spine take place if associated muscles are relaxed; muscle activities or changes in trunk tilt can counter the effects.)

Accordingly, Staffel proposed in 1884 a forward-declining seat surface to open up the hip angle and bring about lordosis in the lumbar area. In the 1960s, a seat pan design with an elevated rear edge became popular in Europe. Since then, Mandal (1975, 1982) and Congleton et al. (1985) again promoted that the whole seat surface slope fore-downward. To prevent the buttocks from sliding down on the forward-declined seat, the seat surface may be shaped to fit the human backside (Congleton), or one may counteract the downward-forward thrust either by bearing down on the feet (Figure 4.18) or by propping the upper shins on special pads (Figures 4.21 and 4.22).

A seat surface that can be tilted throughout the full range (from declined forward, kept flat, to inclined backward) naturally allows the user to assume various curvatures of the lower spinal column, from kyphosis (forward bend) to lordosis (backward bend).

The surface of the seat pan must support the weight of the upper body comfortably and securely. Hard surfaces generate pressure points that can be avoided by suitable upholstery, cushions, or other surface materials that elastically or plastically adjust to body contours.

The only inherent limitation to the size of the seat pan is that it should be so short that the front edge does not press into the leg's sensitive

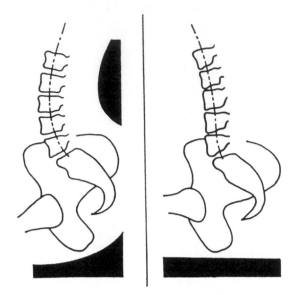

Figure 4.17 Forward rotation of the pelvic bone on its lowest protrusions (the ischial tuberosities) causes forward bending of the lumbar section of the spinal column (lordosis), which can be enhanced by the forward push of a lumbar pad in the backrest. Rearward rotation of the pelvis can flatten the lumbar spine or even bend it backward (kyphosis).

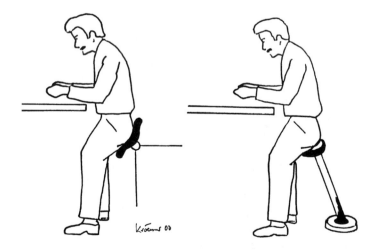

Figure 4.18 Examples of fore-downward tilting surfaces that partially support the body. With permission from Kroemer et al. (2001) © Prentice Hall. All rights reserved.

tissues behind the knees. Usually, the seat pan is between 38 and 42 cm deep and at least 45 cm wide. A well-rounded front edge is mandatory. (This is called a *waterfall* in the trade.) The side and rear borders of the seat pan may be slightly higher than its central part.

The height of the seat pan must be widely adjustable, preferably down to about 37 cm and up to 58 cm to accommodate Western persons with short and long lower legs. It is very important that the person, while seated on the chair, can easily do all adjustments, especially in height and tilt angle.

Figure 4.19 illustrates major dimensions of seat pan and backrest.

Backrest　Two opposing ideas have been promoted vis-à-vis backrests: one advocates not having a backrest at all, so that trunk muscles must remain continually active to keep the upper body in balance. This concept, called active sitting, has not become very popular. Most people think that a backrest is desirable for several reasons. One is that the back support carries some of the weight of the upper body and hence reduces the load that the spinal column must otherwise transmit to the seat pan. A second reason is that a lumbar pad, protruding slightly in the lumbar area, helps to maintain lumbar lordosis, believed to be beneficial. A third, related reason is that leaning against a suitably formed backrest is relaxing.

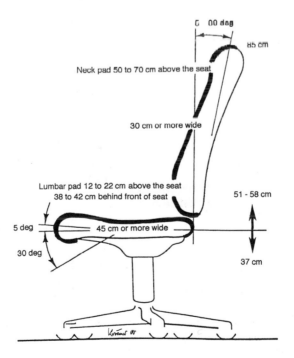

Figure 4.19　Major dimensions of the seat.

Studies have shown the importance of supporting the back by leaning it on a rearward-declined backrest. In some contrast to earlier findings, Andersson et al. (1986, p. 1113) summarized the available literature and concluded: "In a well-designed chair the disc pressure is lower than when standing." Relaxed leaning against a declined backrest is the least stressful sitting posture. This is often freely chosen by persons working in the office if there is a suitable backrest available: "... an impression which many observers have already perceived when visiting offices or workshops with VDT workstations: Most of the operators do not maintain an upright trunk posture. ... In fact, the great majority of the operators lean backwards even if the chairs are not suitable for such a posture" (Grandjean et al., 1984, pp. 100–101).

Of course, the backrest should be shaped to support the back properly: apparently independently from each other, Ridder (1959) in the United States and Grandjean (1963) in Switzerland found in experiments that their subjects preferred similar backrest shapes, as depicted in Figure 4.20. In essence, these shapes follow the curvature of the rear side of the human body. At the bottom, the backrest is concave to provide room for the buttocks, above slightly convex to fill in the lumbar lordosis. Above the lumbar pad, the backrest surface is nearly straight but tilted

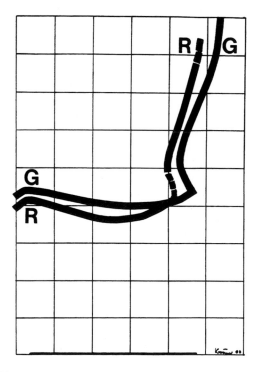

Figure 4.20   Ridder's 1959 and Grandjean's 1963 seat pan and backrest contours.

backward to support the thoracic area, at the top, the backrest is again convex to follow the neck lordosis.

Combined with a suitably formed and upholstered seat pan, this shape has been used successfully for seats in automobiles, aircraft, passenger trains, and for easy chairs. Recall that in the traditional office, the boss enjoyed these first-class shapes while clerical employees had to use simpler designs. The so-called secretarial chairs had a small, often hard-surfaced seat pan and a slightly curved support just for the low back: the more recent task chair is an improved version.

The backrest should be as large as can be accommodated at the work place: this means up to 85 cm high above the seat pan, and up 30 cm wide. To provide support from the head and neck on down to the lumbar region, it is usually shaped to follow the back contours, specifically in the lumbar and the neck regions. Many users appreciate an adjustable pad or an inflatable cushion for supporting the lumbar lordosis. The lumbar pad should be adjustable from about 12 to 22 cm, the cervical pad from 50 to 70 cm above the seat surface (see Figure 4.19).

The angle of the backrest must be easily adjustable while seated. It should range from slightly behind upright (95° from horizontal) to about 30° behind vertical (120°), with further declination for rest and relaxation desirable. Whether or not the seatback angle should be mechanically linked to the seat pan angle is apparently a matter of personal preference.

Armrests   Armrests can provide support for the weight of hands, arms, and even portions of the upper trunk. Thus, armrests can be of help, even if only for short periods of use, when they have a suitable load-bearing surface, best padded. Adjustability in height, width, and possibly direction is desirable. However, armrests can also hinder moving the arm, pulling the seat toward the workstation, or getting in and out of the seat. In these cases, having short armrests, or none, is appropriate.

Footrests   Hassocks, ottomans and footstools have long been popular to put up one's feet, but footrests in the office usually indicate deficient workplace design, like seat pans that cannot be sufficiently lowered for the seated person. If a footrest is used, it should be so high that the sitting person's thighs are nearly horizontal. A footrest should not consist of a single bar or other small surface because this restricts the ability to change the posture of the legs. Instead, the footrest should provide a support surface that is about as large as the total legroom available in the normal work position.

We are all different individuals
The owner of a small suburban consulting firm was perplexed – and, frankly, a little annoyed. He had started his company 3 years ago in the

basement of his house and, after 2 years of 14-hour days and tireless effort, had been able to grow his business enough to lease a posh office suite 3 miles away and hire four associates to help with the work. His most recent employee had joined the company 2 months earlier and her work was excellent. But nonetheless, there was one small aspect about her that was confusing to him – and slightly irritating.

He took great pride in his office and had furnished and appointed it with great care. He was meticulous about choosing not just visually appealing furniture but also items that were properly designed. After all, with a brother-in-law who worked in industrial engineering, he felt he truly understood the importance of well-designed furniture. Accordingly, he had done research, skimped no expense, and purchased the chairs and desks from a reputable office design store. In spite of all of this, however, his newest employee appeared to find her chair less than ideal. Every morning, when she arrived and sat down on her chair, she would place a rolled-up sweater at the small of her back; a few times, she had even set a book along the back of the chair. Last week, she had brought in a special cushion that she now kept on the lumbar support of her chair. When he asked her why, she indicated with a smile that the chair "just wasn't comfortable" for her. How, he thought, could it not be? The chair had been clearly marketed as 'ergonomic' and he had certainly paid extra for this feature. And the other three associates seemed to be perfectly comfortable with their chairs.

Work surface and keyboard support

The height of the workstation depends largely on the activities to be performed with the hands, and how well and exactly the work must be viewed. Thus, the main reference points for ergonomic workstations are the elbow height of the person and the location of the eyes. Both depend on how one sits or stands, upright or slumped, and how one alternates among postures.

The table or other work surface of a sit-down workplace should be adjustable in height between about 53 and 70 cm, even a bit higher for very tall persons, to permit proper hand/arm and eye locations. Often, a keyboard or other input device is placed on the work surface, or connected to it by a tray. A keyboard tray can be useful; especially if the table is a bit high for a person, but it also may reduce the clearance height for the knees. The tray should be large enough for keyboard and trackball or mouse-pad unless these are built into the keyboard.

Sitting and back pain

The posture and movements of the spinal column have been of great concern to physiologists and orthopedists. This is due to the fact that so many persons suffer from annoyance, pain, and disorders in the spinal column, particularly in the low back and in the neck areas (Nelson and Silverstein, 1998). Researchers have explained this by pronouncing that the human body is "not built for long sitting or standing," or not fit because of lack of exercise, or suffering from degeneration, particularly of the intervertebral discs when aging. Physical activities and special exercises can improve fitness. Caution must be applied, however, when selecting exercises: some are appropriate, others are of questionable value, but several are outright dangerous or injurious, as Lee et al. (1992) pointed out.

The most easily applied remedy is to alternate often between walking, standing and sitting. If long-time sitting is required, then the design of the seat and other furniture and equipment is critical; for example, a tall and well shaped backrest that reclines helps to support back and head during work and permits relaxing breaks. While sitting, one should change position often. This can be done purposefully by the person, perhaps with the help of automatic devices that make cushions on the seat or backrest pulsate on and off, or that effect small changes in the angles of seat pan and backrest.

Semi-sitting

Semi-sitting is a posture about halfway between sitting and standing, with some of the upper body weight supported at the buttocks and the rest of the weight transmitted through the legs to the ground. Such semi-seats or stand-seats usually do not have full backrests, if any. Figures 4.21 and 4.22 show several examples of semi-sitting.

Semi-sitting relieves the worker from continuous standing, but is not as supportive as full sitting. Mobility of the trunk is the major advantage of semi-sitting when there is no backrest. One of the great disadvantages of semi-sitting is the tendency to slide off the support surface. This must be counteracted either by fatiguing leg thrust or by pressure against shin pads, which many people find unpleasant or even painful, although some get used to it. With a low semi-seat it can be difficult to move the legs in the confined space between pads and seat as one lowers the body onto the support or arises from it. Some individuals have found semi-sitting acceptable or comfortable but it should not be generally prescribed because most people prefer more conventional seats.

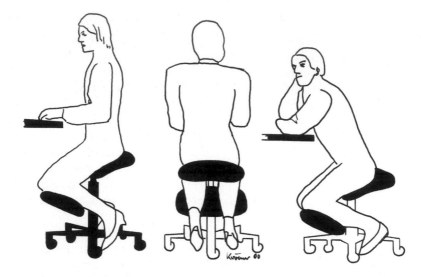

Figure 4.21   Trunk mobility is one advantage of semi-sitting.

Designing the stand-up workstation

Standing up while doing computer work seems like a return to the habits of office clerks common around 1900. Yet, moving about and standing

Figure 4.22   Shin pads help to avoid sliding down a semi-seat.

on one's feet, at least for a period of time, can be a welcome change from sitting, provided that the person does it at his or her own choice. One may choose to stand for reading, writing, or telephoning. Stand-up workstations can use a second computer to alternate work from the sit-down workstation for a while – or one may simply move a handheld or laptop computer from one workstation to the other. Some people prefer standing and walking altogether to sitting in the office.

Stand-up workstations should be adjustable to have the working area used for writing or computer inputs at approximately elbow height when standing, between 0.9 and 1.2 m above the floor. As in the sit-down workstation, the display should be located close to the other visual targets and directly behind the keyboard. If the work surface is used for reading or writing, it may slope down slightly toward the person. A footrest at about two-thirds knee height (approximately 0.3 m) allows the person to prop one foot up on it temporarily. This brings about welcome changes in pelvis rotation and spine curvature.

Non-resilient floors, such as those made of concrete, can be hard on people's feet, legs, and backs. Carpets, elastic floor mats and soft-soled shoes can reduce strain. Appropriate friction between soles and the walkway surface helps to avoid slips and falls.

The Jungle Gym – or the importance of moving around

Cheryl, the manager of a corporate secretarial pool, was having difficulty placing John, one of the secretaries; although John's work was more than satisfactory, his supervisor in the most recent job rotation called him restless and said that John's seeming inability to sit still made his colleagues uneasy. Perplexed by this comment – Cheryl had always thought highly of John's work skills – she asked the rotation supervisor to elaborate on John's evaluation. The supervisor explained that the secretary appeared to move around a great deal during his work, even though the duties themselves were largely sedentary in nature. Specifically, most of John's tasks involved using a computer to type in and print out various correspondences and reports. John's output was apparently perfectly acceptable, but he seemed to dislike sitting 'normally', as the supervisor put it. Generally, secretaries would remain seated at their keyboards for almost their entire 8-hour shift, taking breaks only for lunch or to get coffee. John, however, moved around: he would sit at his work station sometimes, but would kneel in front of his computer at other times, even work standing up on occasion; additionally, he arose frequently to stretch and take quick strolls. "John doesn't need a chair," the supervisor commented, "he needs a jungle gym."

Designing the home office

Nearly everything said above applies to the home office. Yet, in arranging our home office we are inclined to disregard the ballast of old habits and conventions about how to sit at work that governs us in the company office (Crantz, 2000).

If you work only for short periods of time in your office at home, then the old dining table and the odd kitchen chair probably will not harm you. But as soon as you get serious about using your home office, working in it for hours, you should become very conscious about the working conditions there. Equip your office with carefully selected furniture, where the components of the workstation fit each other well, and, most importantly, fit you well; do not do as shown in Figure 4.23.

Select an easy chair, a comfortable office chair (even if it is expensive) or a semi-sit perch or a kneeling chair – whatever feels good to you – that supports your body well over long periods of time. Perhaps you want to work standing up, at least when you read or make phone calls, for example. Nobody dictates how to set up your own home office: you can and should do what suits you.

> John is a 46-year old salesman who works half of the time out of his own office in a suburb of Washington. He connects with his customers via a telephone with head-mounted speaker and mike. In the morning, some clients may just barely hear a muffled thump-thump-thump when they speak with him on the phone because he jogs, slowly, on his treadmill. He has mounted a support surface for his notebook computer on the handlebars at the front end of the treadmill so that he can use the computer while walking. For more complex input tasks, he stops the treadmill and rests in his lounge chair.

Do not fall for furnishings that are un-ergonomic, such as shelving units that make you put the monitor up high, on a shelf much above the keyboard, or surfaces that do not provide sufficient space for both the keyboard and the mouse pad. This is your own workspace, put together for your comfort and ease at work, and it does not have to be similar to anybody else's set-up – nor does it have to be expensive, because some simple furniture on the market is well designed.

Get a quality computer with an up-to-date display and with suitable input devices. Select a keyboard that feels comfortable to you (Chapter 5), but consider voice input that might serve you well. If you travel with your handheld or laptop computer, consider a docking station at home. Do you want to use a laptop computer in your office as well?

Select a room with good lighting that is separate, quiet and well heated and cooled (Chapters 7, 8, and 9). You will probably spend more time in

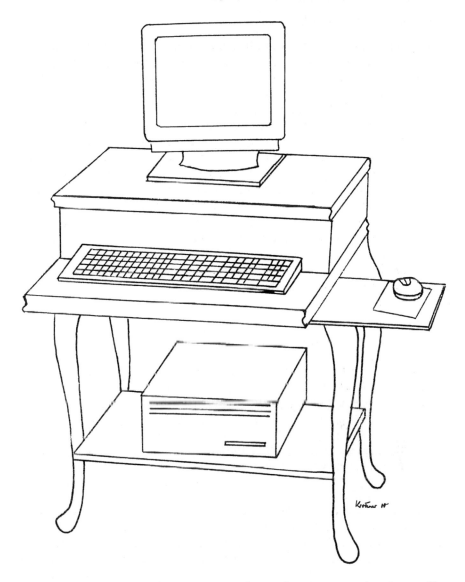

Figure 4.23   About everything is wrong in this workstation meant for a home office: the display is too high in the top shelf. The keyboard takes too much space and a sharp edge of the table is likely to cut into the wrists. The mouse-pad is too far to the side. The bottom shelf leaves no space for the legs.

your home office than you expected, and your wellbeing is worth the effort and money that you spend.

Mike, a young engineering school student at a prominent Southeastern university, is home for spring break. While he's been away at school for a year, his mother and sister have turned their passion for jams and jellies into a small home-based business. Thanks to a high-quality product, a devoted local following, and recent inroads in distribution through a regional grocery store chain, "Marge's Magical Marmalades" is growing. In fact, Mike's mother Marge has just acquired two computers for their basement home office; one for her use, and one for her daughter's. They plan to use the computers for inventory tracking, logging orders, customer correspondence, billing, and financial reporting.

Of late, Mike's mother has been complaining of back pain, and his sister has mentioned aches in her neck. Having just taken a course in ergonomics, he hypothesizes that there may be a connection between their business success and their physical discomfort. He asks for a tour of the basement office. Thrilled about her son's interest, Marge proudly shows him around her hastily refurbished basement, and he slowly looks around and takes in the view. Her beloved new computer is perched on her L-shaped desk, the monitor glowing brightly from its position on the long side of the desk. The keyboard rests on the shorter portion of the desk, separate from and at right angle to the monitor. Marge's chair fits cleanly under the shorter portion of the desk but is too tall to slide under the longer portion; which explains why she had placed the keyboard away from the monitor. The chair, unfortunately, does not have height adjustment capabilities. Mike immediately realizes that this configuration forces his mother to twist her body whenever she used her computer, turning one way to access the keyboard, and the other to watch the monitor. "Hmmm," he mumbles. Then, he turns to his sister's work station. His mother's desk is the only true desk in the cramped office space; since there is no room for another table, his sister's computer sits on top of a three-drawer beige filing cabinet in the corner. To use it while seated, she can keep the keyboard on her lap, but must place her legs on either side of the cabinet and crane her head upward to view the monitor, which is located about a foot higher than her eyes. "Uhhuhh," he mutters.

Marge turns to her son. "Hmmm?" she says, "Uhhuhh? What does that mean?" "Well", he replies, "I think I know what the problem is with your back, Mom, and my sister's neck.".

A few more words about furniture adjustments

Traditionally, much emphasis has been placed on the need for easy adjustments especially of chairs, of support surfaces for the keyboard, and of the display in height, distance and angle – and we have repeated these demands in our text above. Yet, many formal studies (for example by Vitalis et al., 2000) supported our personal experiences that existing adjustment features are seldom used. One conclusion is that the adjustment means are still too complicated. Another related inference is that all adjustments should be automatic, not requiring any conscious action by the user at all. For example, the seat should simply follow all movements of the sitting person, and support the body throughout. Yet, one unresolved problem is that even as we change our sitting habits, keyboard and display stay in place if they are not coupled with the seat. This is true for a conventional desk top computer system but not with a hand held or laptop computer. Let us hope then that engineers will solve this problem with technological progress, e.g., a virtual or head-coupled display and voice input to the computer would set us free (Weiland et al., 2000).

Fitting it all together

A bad or mediocre office is not instantly converted into a good one by changing to a different computer, or by simply acquiring a better chair. All the components, equipment and furniture (see Figure 4.24) and lighting and climate must fit each other, and the person in the office must be willing and able to take advantage of all the offered possibilities. Equipment change often brings change in mental attitude. This applies not only to the working person, but also to management: the old fashioned expectation that the postures of all the office workers and the look of all the office furniture be the same is not reasonable and is in fact damaging and counter-productive.

ERGONOMIC DESIGN RECOMMENDATIONS

Neither theories nor practical experiences endorse the idea of one single proper, healthy, comfortable sitting position. Instead, many motions and postures may be subjectively comfortable (healthy, preferred, suitable) for short periods of time, depending on one's body, preferences, and work activities.

The traditional postulate that everybody should sit upright is thus abolished, and furniture should be designed for free-flowing motion. Changing from one posture to another one, in fact moving freely among all the comfortable poses, is advisable. Motion, change, variation and adjustment to fit the individual are central to well-being.

Consequently, furniture should allow for body movements among various postures. Accordingly, furniture should adjust automatically, at least be easily adjustable in its main features, especially in seat height and seat pan angle and backrest position. The entire compu-

In 1999, Liberty Mutual completed an investigation of the impact of flexible office workspaces and ergonomic training on employee health and performance. 20 office workers moved into new adjustable workspaces while another 20 occupied new but corporate-specific workplaces. All 40 people received training in ergonomics. The expectation was that, by giving employees more control over their environment and a better understanding of ergonomic principles, performance would improve and health problems diminish. The results of the 18-month study confirmed the expected: combined with ergonomic training, the flexible workspace increased individual performance and group collaboration. This was accompanied by a nearly one-third reduction in back pain and a two-thirds reduction in upper limb pain among the employees who had more control over their environment.

ter workstation should permit easy variations, for example in the location (especially height) of the input devices and height and distance of the display.

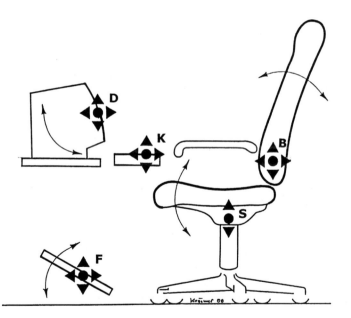

Figure 4.24   All components of the workstation must fit each other: Seat pan S and Backrest B, Keyboard support K and Display location D, and a Footrest F may be needed.

This feels good

We should re-design our computer systems and how we interact with them to better fit our abilities (see Chapter 5). But even if we do not change our office tools and habits fundamentally, we still can make our work as easy as possible by doing the following:

- Place all the things you must operate with your hands (keyboard, mouse, trackball, pen, paper, telephone)

  - directly in front of you,
  - at elbow height,
  - within easy reach.

- Place all the things you must clearly see – display, source document, writing pad, template, keyboard (see also Chapter 5)

  - directly in front of you, at your best viewing/reading distance (currently, your computer display is probably too far away from you)

- Place the display

  - low behind your keyboard (do not put a tall tilt stand, or the CPU unit, under the computer display).

- Sit on a seat designed so that you can change your posture frequently. If long-time sitting is required, then a tall backrest that can recline helps support back and head. (At least, this allows you to quickly take a relaxing break.)
- Change your body position often. Change helps avoid continued compression of tissues, especially of the spinal column, facilitates blood circulation and counteracts muscular fatigue – and it breaks boredom.
- Support your arms and hands by resting them, as often as feasible, on soft arm rests attached to the seat, and on padded wrist rests at the keyboard – but avoid hard surfaces and, worse, rigid corners and edges that compress the skin tissues.
- Keep the shoulders relaxed, the upper arms hanging down, the forearms horizontal and the wrists straight. To achieve this, all elements of your workstation (chair, computer on its support, table and desk) must be carefully arranged in concert with each other.
- Follow recommendations in ANSI/HFES 100.

What to do if you are not comfortable

If your eyes are tired, feel teary, or hurt

- Place all the things you must clearly see (display, source document, writing pad, template, keyboard).

  - directly in front of you, to the best viewing/reading distance (which is probably shorter than what you presently have).

- Place the display

  - low behind your keyboard (do not use a tall tilt stand, or the CPU unit, under the monitor).

- Make sure you do not have light (from a window or from lamps) reflected in the display, or shining directly into your eyes (see Chapter 7).
- Talk with your supervisor and, if the condition does not go away, have your eyes checked by a specialist (ophthalmologist or optometrist).

If your back hurts

- Take a break at least every 30 min; walk and move your body.
- Make sure that you lean against the backrest of your seat.
- Get a seat that fits your body and accommodates your sitting habits better than your current chair.
- Place all the things you must clearly see (display, source document, writing pad, template, keyboard)

  - directly in front of you, at your best viewing/reading distance (which is probably shorter than what you presently have).

- Place the display

  - low behind your keyboard (do not use a tall tilt stand, or the CPU unit, under the monitor).

- Place all the things you must operate with your hands (keyboard, mouse, trackball, pen, paper, telephone)

  - directly in front of you,
  - at elbow height,
  - within easy reach.

- Talk with your supervisor and get a medical evaluation if the condition does not abate.

If your neck hurts

- Take a break at least every 30 min; walk and move your body.
- Place all the things you must clearly see (display, source document, writing pad, template, keyboard)

  - directly in front of you, tat your best viewing/reading distance (which is probably shorter than what you presently have).

- Place the display

  - low behind your keyboard (do not use a tall tilt stand, or the CPU unit, under the monitor).

- Place all the things you must operate with your hands (keyboard, mouse, trackball, pen, paper, telephone)

  - directly in front of you,
  - at elbow height,
  - within easy reach.

- Talk with your supervisor and get a medical evaluation if the condition does not go away.

If your shoulder hurts

- Take a break at least every 30 min; walk and move your body.
- Put the mouse or trackball next to the keyboard, all at elbow height.
- Operate the mouse or other input device with the other hand (yes, you can do so well after just a few minutes).
- Use armrest and wrist rest often.
- Talk with your supervisor and get a medical evaluation if the condition does not go away.

If your wrist or hand hurts

- Take a break at least every 30 min; walk and move your body.
- Make sure that your wrist remains straight while working the keyboard or other input device.
- Strike keys very lightly.
- Use armrest and wrist rest often.
- Put the mouse or trackball next to the keyboard.

- Talk with your supervisor and get a medical evaluation if the condition does not go away.

If your leg hurts

- Take a break at least every 30 min; walk and move your body.
- Make sure that you have ample room at your workstation to position and move your feet freely.
- If the front portion of your seat presses on the underside of your thighs

  - lower the seat (most likely, you also must lower keyboard and monitor, table or desk or other work surface accordingly),
  - get another seat that has a soft waterfall shape at its front,
  - use a wide and deep footrest.

- Talk with your supervisor and get a medical evaluation if the condition does not improve.

Acknowledgement

Our thanks to Mrs. M. J. Kennedy for having allowed us to use her drawing in Figure 4.1 for some 25 years.

References

Aaras, A., Fostervold, K. I., Ro, O., Thoresen, M. and Larsen, S. (1997). Postural Loading During VDU Work: A Comparison Between Various Work Postures. *Ergonomics 40*, 1255–1268.

Ankrum, D.R. and Nemeth, K.J. (1995). Posture, Comfort, and Monitor Placement. *Ergonomics in Design*, April Issue, 7–9.

ANSI/HFS 100 (1988). *American National Standard for Human Factors Engineering of Visual Display Terminal Workstations*. Santa Monica, CA: Human Factors Society.

ANSI/HFES 100 (2001). *US National Standard for Human Factors Engineering of Computer Workstations*. Santa Monica, CA: Human Factors and Ergonomics Society, in press.

Andersson, B. J. G. and Oertengren, R. (1974). Lumbar Disc Pressure and Myoelectric Back Muscle Activity During Sitting. II. Studies on an Office Chair. *Scandanavian Journal of Rehabilitation Medicine 6*, 115–121.

Andersson, B. J. G., Oertengren, R., Nachemson, A. and Elfstroem, G. (1974). Lumbar Disc Pressure and Myoelectric Back Muscle Activity During Sitting. I. Studies on an Experimental Chair. *Scandanavian Journal of Rehabilitation Medicine 6*, 104–114.

Andersson, B. J. G., Schultz, A. B. and Oertengren, R. (1986). Trunk Muscle Forces During Desk Work. *Ergonomics 29,* 1113–1127.

Bauer, W. and Wittig, T. (1998). Influence of Screen and Copy Holder Positions on Head Posture Muscle Activity and User Judgment. *Applied Ergonomics 29*, 185–192.

Bendix, T., Poulsen, V., Klausen, K. and Jesnen, C. V. (1996). What Does a Backrest Actually Do to the Lumbar Spine? *Ergonomics 39*, 533–542.

Bonne, A. J. (1969). On the Shape of the Human Vertebral Column. *Acta Orthopaed. Belg. 35*(3–4), 567–583.

Booth–Jones, A. D., Lemasters, G. K., Succop P., Atterbury, M. R. and Bhattacharya, A. (1998). Reliability of Questionnaire Information Measuring Musculoskeletal Symptoms and Work Histories. *American Industrial Hygiene Association Journal 59*, 20–24.

Bradford, P. and Byrne, W. (1978). *Chair*. New York: Crowell.

Crantz, G. (2000). Computer Workstations of the Future, in *Proceedings of the XIVth Triennial Congress of the International Ergonomics Association and 44th Annual Meeting of the Human Factors and Ergonomics Society*. Santa Monica, CA: Human Factors and Ergonomics Society, 6708–6711.

Congleton, J. J., Ayoub, M. M. and Smith, J. L. (1985). The Design and Evaluation of the Neutral Posture Chair for Surgeons. *Human Factors 27*, 589–600.

Corlett, E. N. and Bishop, R. P. (1976). A Technique for Assessing Postural Discomfort. *Ergonomics 19*, 175–182.

Dainoff, M. J. (1999). Ergonomics of Seating and Chairs, in W. Karwowski and W. S. Marras (Eds.), *The Occupational Ergonomics Handbook*. Boca Raton, FL: CRC Press, 1761–1778.

Dickinson, C. E., Campion, K., Foster, A. F., Newman, S. J., O'Rourke, A. M. T. and Thomas, P. G. (1992). Questionnaire Development: An Examination of the Nordic Musculoskeletal Questionnaire. *Applied Ergonomics 23*, 197–201.

Donkin, S. W. (1987). *Sitting on the Job*. Lincoln, NE: Parallel Integration.

Dowell, W. R., Price, J. M. and Gscheidle, G. M. (1997). The Effect of VDT Screen Distance on Seated Posture, in *Proceedings of the Human Factors and Ergonomics Society 41st Annual Meeting*. Santa Monica, CA: Human Factors and Ergonomics Society, 505–508.

Genaidy, A. M. and Karwowski, W. (1993). The Effects of Neutral Posture Deviations on Perceived Joint Discomfort Ratings in Sitting and Standing Postures. *Ergonomics 36*, 785–792.

Grandjean, E. (1963). *Physiological Design of Work*. Thun: Ott (in German).

Grandjean, E. (1987). *Ergonomics in Computerized Offices*. London: Taylor & Francis.

Grandjean, E., Huenting, W. and Nishiyama, K. (1984). Preferred VDT Workstation Settings, Body Postures and Physical Impairments. *Applied Ergonomics 15*, 99–104.

Hagberg, M. and Rempel, D. (1997). Work-related Disorders and the Operation of Computer VDTs, in M. Helander, T. K. Landauer and P. Prabhu (Eds.) *Handbook of Human-Computer Interaction* (2nd ed.). Amsterdam: Elsevier, 1415–1429.

Hamilton, N. (1996). Source Document Position as it Affects Head Position and Neck Muscle Tension. *Ergonomics 39*, 593–610.

Halpern, C. A. and Davis, P. J. (1993). An Evaluation of Workstation Adjustment and Musculoskeletal Discomfort, in *Proceedings of the Human Factors and Ergonomics Society 37th Annual Meeting*. Santa Monica, CA: Human Factors and Ergonomics Society, 817–821.

Helander, M. G. and Zhang, L. (1997). Field Studies of Comfort and Discomfort in Sitting. *Ergonomics 401*, 895–915.

Hill, S. G., and Kroemer, K. H. E. (1986). Preferred Declination and the Line of Sight. *Human Factors 28*, 127–134.

Kroemer, K.H.E. (1985). Office Ergonomics: Work Station Dimensions. Chapter 18 in D.C. Alexander and B.M. Pulat (Eds.), *Industrial Ergonomics* (187–201). Norcross, GA: Institute of Industrial Engineers.

Kroemer, K.H.E., Kroemer, H.J. and Kroemer-Elbert, K.E. (1997). *Engineering Physiology* (3rd ed.). New York: Van Nostrand Reinhold–Wiley.

Kroemer, K. H. E., Kroemer, H. B. and Kroemer-Elbert, K. E. (2001). *Ergonomics: How to Design for Ease and Efficiency* (2nd ed.). Upper Saddle River, NJ: Prentice Hall.

Kroemer, K.H.E. and Grandjean, E. (1997). *Fittting the Task to the Human* (5th ed.). London: Taylor & Francis.

Kuorinka, I., Jonsson, B., Kilbom, A., Vinterberg, H., Biering-Sorensen, F., Andersson, G. and Jorgensen, K. (1987). Standardized Nordic Questionnaires for the Analysis of Musculoskeletal Symptoms. *Applied Ergonomics 18*, 233–237.

Lee, N., Swanson, N., Sauter, S., Wickstrom, R., Waikar, A. and Mangum, M. (1992). A Review of Physical Exercises Recommended for VDT Operators. *Applied Ergonomics 23*, 387–408.

Lehmann, G. (1962). *Praktische Arbeitsphysiologie* (2nd ed.). Stuttgart: Thieme. (In German.)

Liberty Mutual Research Center for Safety and Health (1999). *From Research to Reality. (Annual Report)*. Hopkinton, MA: Liberty Mutual Research Center for Safety and Health, 14.

Mandal, A. C. (1975). Work-Chair with Tilting Seat. *Lancet i*, 642–643.

Mandal, A. C. (1982). The Correct Height of School Furniture. *Human Factors 24*, 257–269.

Merrill, B. A. (1995). Contributions of Poor Movement Strategies to CTD Solutions or Faulty Movements in the Human House, in *Proceedings of the Silicon Valley Ergonomics Conference & Exposition, ErgoCon 95*. San Jose, CA: San Jose State University, 222–228.

Michel, D. P. and Helander, M. G. (1994). Effects of Two Types of Chairs on Stature Change and Comfort For Individuals with Healthy and Herniated Disks. *Ergonomics 37*, 1231–1244.

Nag, P. K., Chintharia, S. and Nag, A. (1986). EMG Analysis of Sitting Work Postures in Women. *Applied Ergonomics 17*, 195–197.

Nelson, N. A. and Silverstein, B. A. (1998). Workplace Changes Associated with a Reduction in Musculoskeletal Symptoms in Office Workers. *Human Factors 40*, 337–350.

Pheasant, S. (1996). *Bodyspace* (2nd. ed.). London: Taylor & Francis.

Ridder, C. A. (1959). *Basic Design Measurements for Sitting* (Bulletin 616, Agricultural Experiment Station). Fayetteville, AR: University of Arkansas.

Shackel, B., Chidsey, K. D. and Shipley, P. (1969). The Assessment of Chair Comfort. *Ergonomics 12*, 169–306.

Staffel, F. (1884). On the Hygiene of Sitting. *Zbl. Allgemeine Gesundheitspflege 3*, 403–421 (in German).

Staffel, F. (1889). *The Types of Human Postures and Their Relations to Deformations of the Spine*. Wiesbaden: Bergmann (in German).

Tenner, E. (1997). How the Chair Conquered the World. *Wilson Quarterly 21*, 64–70.

Tenner, E. (1997). The Life of Chairs. *Harvard Magazine January/February*, 47–53.

Turville, K. L., Psihogios, J. P. and Mirka, G. A. (1998). The Effects of Video Display Terminal Height on the Operator: A Comparison of the 15 and 40 Recommendations. *Applied Ergonomics 29*, 239–246.

Van der Grinten, M. P. and Smitt, P. (1992). Development of a Practical Method for Measuring Body Part Discomfort, in S. Kumar (Ed.), *Advances in Industrial Ergonomics and Safety IV*. London: Taylor & Francis, 311–318.

Veiersted, K. B. (1998). Arm Pain related to Mechanical Exposure of VDU Use. A Review of the Epidemiology, in S. Kumar (Ed.), *Advances in Occupational Ergonomics and Safety*. Amsterdam: IOS Press, 453–455.

Villanueva, M. B. G., Sotoyama, M., Jonai, H., Takeuchi, Y. and Saito, S. (1996). Adjustments of Posture and Viewing Parameters of the Eye to Changes in the Screen Height of the Visual Display Terminal. *Ergonomics 39*, 933–945.

Vink, P. and Kompier, M. A. J. (1997). Improving Office Work: a Participatory Ergonomics Experiment in a Naturalistic Setting. *Ergonomics 40*, 435–449.

Vitalis, A., Marmaras, N., Poulakis, G. and Legg, S. (2000). Please Be Seated, in *Proceedings of the XIVth Triennial Congress of the International Ergonomics Association and 44th Annual Meeting of the Human Factors and Ergonomics Society*. Santa Monica, CA: Human Factors and Ergonomics Society, 6436–6446.

Weiland, W. J., Zachary, W. W. and Stokes, J. M. (2000). Personal Wearable Computer Systems, in *Proceedings of the XIVth Triennal Congress of the International Ergonomics Association and 44th Annual Meeting of the Human Factors and Ergonomics Society*. Santa Monica, CA: Human Factors and Ergonomics Society, 6712–6715.

Williams, M. M., Hawley, J. A., McKenzie, R. A. and van Wijmen, P. M. (1991). A Comparison of the Effects of Two Sitting Postures on Back and Referred Pain. *Spine 16*, 1185–1191.

Wright, W. C. (1993). *Diseases of Workers. Translation of Bernadino Ramazzini's 1713 De Morbis Articum*. Thunder Bay, ON: OH&S Press.

Zacharkow, D. (1988). *Posture: Sitting, Standing, Chair Design and Exercise*. Springfield, IL: Thomas.

5 Keyboarding and other manual tasks

Overview

Writing words and numbers by hand was the typical task in an old-fashioned office. Maintaining a firm grasp on a pen and guiding it ceaselessly in finely controlled motions over paper generates a repetitive strain of the musculoskeletal system that can cause, then as today, a repetitive strain injury, called RSI for short. This was simply called scribe's or writer's cramp 300 years ago. Keyboarding has replaced much of the handwriting in the computerized office and has created new ergonomic tasks and challenges in the process.

The human hand: what it can and cannot do

The hand is a remarkably versatile part of our body. It can touch and grip, manipulate forcefully and with fine control, hold delicately and move energetically. It is enduring and sturdy but can be injured by a sudden blow or cut. More relevantly to us in the office, it can be damaged by often-repeated small impacts such as from external vibrations or by internal efforts to press keys.

Some of us are more susceptible to such repetition motion overload, and some of our jobs require more repetitive motions a day than others. Biomechanically, our hands are meant to do many different activities, with breaks between the chores, but they are not equipped to do the same action over and over for hours on end, which is inherent in continuous handwriting or keyboarding.

Feeling comfortable

For our wellbeing and comfort, our office tasks should be varied over the day, require miscellaneous efforts, and use our manifold mental and manual capabilities. Typing all day, for example, can be both boring mentally and stressful physically. Therefore, we should ergonomically re-design our office equipment (especially keyboards) while also re-evaluating our other office tasks to fully utilize our capabilities without over-taxing them.

> You may skip the following part and go directly to the Ergonomic Design Recommendations at the end of this chapter – or you can get detailed background information by reading the following text.

What, exactly, are the manual tasks in the office?

Hand and wrist

Our hands are solidly built on 27 bones: 8 carpals at the base next to the wrist joint, 5 metacarpals ending in the knuckles and 14 phalanges of the thumb and four fingers. Figure 5.1 shows the bones of the hand. There is little motion among the carpal bones, but the sections of the digits can bend and extend by rotating the bony links in their joints. Their movement is controlled by the pull of tendons. These are cable-like tissues, attached to the bones of the digits and, on the other end, to muscles, most

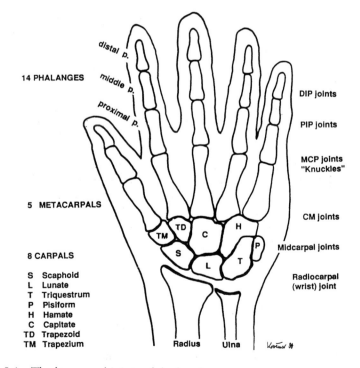

Figure 5.1   The bones and joints of the hand.

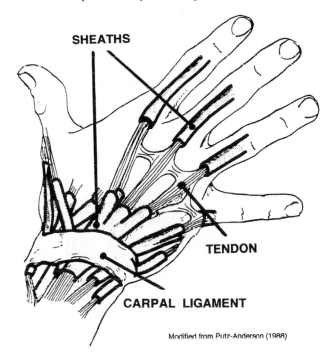

SHEATHS

TENDON

CARPAL LIGAMENT

Modified from Putz-Anderson (1988)

Figure 5.2   The back of the hand with the digit tendons that extend (straighten) the
fingers and thumb. (Modified and with permission from Putz-Anderson
(1988) © Taylor & Francis Ltd.)

of which are located in the forearm. A shortening of a muscle pulls its
tendon by the same distance. Figure 5.2 shows the back of the hand with
its tendons that extend (straighten) the five digits, four fingers and the
thumb. The cable-like tendons are, over much of their lengths, encapsu-
lated in sheaths.

The tendons coming from the muscles in the forearm must cross the
wrist joint and then pass through a tight passage. The carpal bones form
a channel that is covered by the transverse ligament, which makes it the
carpal channel or tunnel. Figure 5.3 shows schematically how blood
vessels, nerves and all the flexor tendons thread through the narrow
opening of the carpal tunnel. The extensor tendons run on the backside
of the carpal bones.

In a digit, the flexor and extensor tendons lie, within their sheaths,
along the palm and the back surfaces of the bones of the phalanges. The
tunnel-like sheaths keep them in place. Figure 5.4 sketches sheaths and
flexor tendons for three fingers and the thumb, seen from the palmar side
of the hand.

Figure 5.5 provides a side view of the flexor tendon of a straight finger.
Tough ring-like and crossed ligaments form part of the tube-shaped

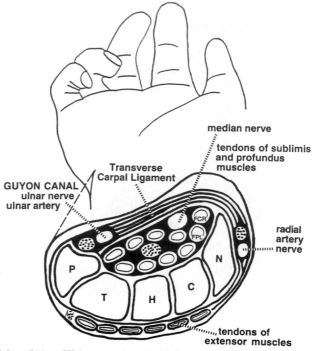

FPL flexor pollicis longus FCR flexor carpi radialis P pisiform T triquetrum H hamate C capitate N navicular

Figure 5.3   Cross-section of the right hand, about 3 cm distal from the wrist joint. This simplified view shows carpal bones and the carpal tunnel with blood vessels, nerves and digit flexor tendons crowded into its tight space. (Modified and with permission from Kroemer et al. (2001) © Prentice Hall.)

sheath that keeps the tendon in place from the palm area of the hand, along the finger's length, until it attaches to surface of the distal phalanx. Shortening the muscle at the proximal end of the tendon pulls it, inside the sheath, until the finger flexes in its joints. The annular and cruciform pulleys of the sheath bend the digit under the pull force of the tendon, as shown in the simple biomechanical model of Figure 5.6. Straightening of the finger is done by the opposing action of an extensor tendon, pulled upon by its specific muscle. The movements of tendons can be very large, such as the full 9 or more cm that a tendon slides within its sheath (Treaster and Marras, 2000).

Tension forces within a tendon, and within a tendon-sheath pulley, can be multiples of the force applied with a fingertip. When the wrist joint is bent, the tendons crossing it must bend also. This adds more strain to the tissues of tendons and sheaths. Frequent movement can lead to wear and tear of the tendons and their sheaths and consequently to repetitive strain injury.

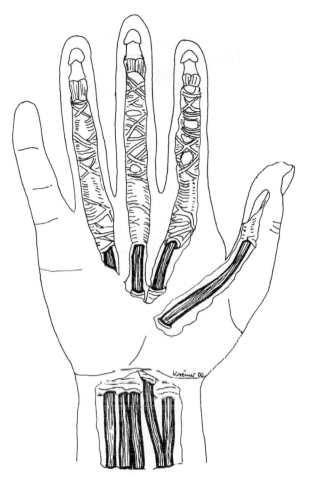

Figure 5.4 Tendons that bend digits of the hand are guided by tube-like sheaths through the wrist area and along the digits.

Repetitive strain from keyboarding

Extensive keyboard operation has long been known to be mentally and physically stressful. The highly repetitive activity of pressing keys with the digits of the hand is especially stressful and often overexerts the human motor system. Aches and pains in the hand/wrist/forearm region commonly occur in typists and players of musical instruments; tendinitis, tenosynovitis and carpal tunnel syndrome are common medical diagnoses. Waves of health complaints have been reported among keyboard users: first in the 1960s and 1970s in Japan, then in the early 1980s in Australia, followed by outbreaks among newspaper reporters and other keyboarders in the USA in the 1990s. Injury often occurs in the carpal

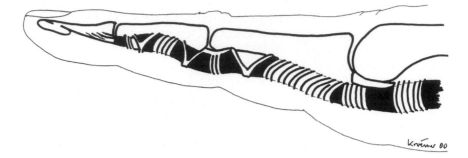

Figure 5.5   Ring-like and cruciform ligamentous pulleys are part of the sheath keeping a flexor tendon close to the bones of a finger.

tunnel, with syndromes related to increased pressure in the carpal tunnel, reduced blood supply and reduced functioning of the median nerve. See Chapter 6 for more on this topic.

The overexertions, especially of the tendon/sheath unit, often near the wrist area, are basically a mechanical overuse problem. The tendon is akin to a cable under tension rubbing vigorously against the inner surface of its sleeve (the tendon sheath) with lubrication (by synovial fluid) that can fail with overuse. The condition is worsened if tension and trans-

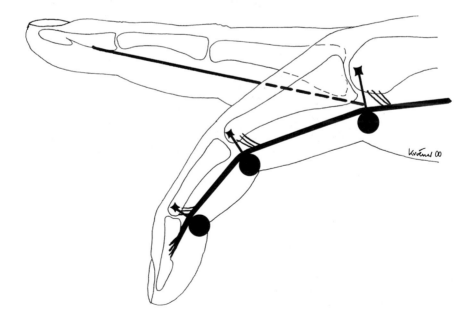

Figure 5.6   A simple biomechanical model of how pulling on a tendon bends a finger through the effects of pulleys.

mitted force are increased, the frequency elevated, and the tendon bent due to a flexed, extended, or laterally deviated wrist. Biomechanically similar conditions can occur on the shop floor in manufacture or assembly, on the construction site in carpentry or bricklaying, or in the office while keying or using the computer's mouse, called mousing (Armstrong and Lackey, 1994; Kroemer, 1992; Kuorinka and Forcier, 1995; Moon and Sauter, 1996; Putz-Anderson, 1988; Rempel et al., 1992; Treaster and Marras, 2000). As Kroemer et al. (2001) explain, the biomechanical and pathological events described above are well understood, but the consequences – injuries – are more prevalent than ever. In the workplace, there is a large and pressing need for more suitable design of manual tasks, and relatedly, of equipment, especially for computer data input.

There is one thing that the keyboard user can do to reduce the incidence of RSIs independently of any equipment design: be aware of one's own working habits.

- Press the keys (and the mouse) lightly; do not pound on keys.
- Keep your shoulders relaxed, let the upper arms hang down; do not tense your muscles.
- Rest your wrists on padded supports between typing or mousing.
- Interrupt your keying and mousing frequently; many short breaks are better than a few long pauses, even if all add up to the same time off.
- Keep your wrists straight. Bending them sideways, or up or down (as in Figure 5.7), impedes the movements of the tendons that control your fingers and thumb.

Figure 5.7   Elevating the wrist, or lowering the wrist, impedes the smooth functioning of the tendons that move the hand's thumb and fingers and may contribute to carpal tunnel syndrome or other overexertion injuries. (Modified from a sketch provided courtesy of Herman Miller, Inc.) Keep the wrists straight while keyboarding – and consider putting the monitor down, directly behind the keyboard, as discussed in Chapter 7.

Keyboard design

The origin of the QWERTY keyboard

Numerous attempts to design writing machines began before the 1870s. Would-be inventors created a multitude of different keyboards, some similar to the long black and white keys on the piano, some featuring double or triple rows of button-like keys, some showing keys arranged in concave or convex circle segments (Adler, 1997; Herkimer, 1923; Martin, 1949). In most cases, the designation of keys to certain letters or numerals or other signs was not made clear.

C. Latham Sholes was the first inventor to successfully design, produce and market a typewriter. With it came a special keyboard. His Type-Writing Machine, patented on August 27, 1878 (US Patent 207,559) shows a keyboard with four rows of a total of 44 keys (see Figure 5.8). The keys on the third row (counted from the operator) are labeled, from the left, QWERTY. This arrangement of those six keys is often used nowadays as a short label for the arrangement of all the letter (alpha or alphabetic) keys. Remarkably, the current QWERTY keyboard is still arranged essentially as Sholes did it, with only the positions of the letters Z and X exchanged and the letter M is moved by one row.

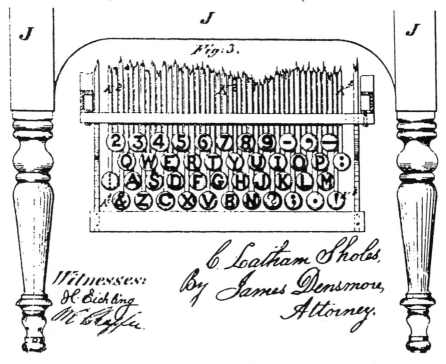

Figure 5.8   Sholes' 1878 QWERTY keyboard. Excerpt from Figure 5.3 in his Letter's Patent 207,559 dated August 27, 1878.

How Sholes decided on this arrangement is not known. He had previously obtained a series of patents on typing machines (in 1868, 79,265 and 79,868; in 1876, 182,511; in 1878, 199,382, 200,351, 207,557 and 207,558), several co-authored with other inventors. In Patent 207,559, Sholes made 14 specific technical claims, but none refers to the key selection or layout. One drawing in this patent shows a frontal view of the invention with four staggered, horizontal rows of keys, with the row farthest from the operator the highest. Another drawing depicts a top view of the four straight rows, shown in Figure 5.8, where the letters QWERTY start the third row of the keyboard. Each of the four rows of keys consists of 11 keys.

Sholes' first 1868 patent (79,265) shows 10 short keys above 11 longer ones, indeed, as he wrote, "similar to the key-board of a piano." In his next 1868 patent (79,868), all the keys, 36 in total, have been flattened into the same plane but they alternate in length. The keys on the left side carry inscribed numbers while the others show letters in alphabetical order. The patent's text provides no explicit explanation for this key layout. His 1876 patent and the first four patents of 1878 all have a very different key pattern, three straight rows of button-like keys affixed to lever-type bars. The buttons show no letters or numbers, and no explanations are given anywhere in the patent texts.

Lacking any statements by Sholes or his co-inventors, one can only observe that the 1878 QWERTY layout shows some remnants of an alphabetic arrangement. Sholes was a printer by trade, and so one can also surmise similarities to the arrangement of the printer's type case in which pieces presumably were assorted according to convenience of use and not according to the alphabet.

Another possible reason for the arrangement of the letters and keys may have been the intent to avoid the type bars (of the then-used mechanical typewriters) colliding or sticking together when neighboring bars were activated in a quick sequence. It is likely that this led to a separation of certain bars and keys on the keyboard; however, there is no contemporaneous evidence for this.

Thus, the reasons for Sholes' layout of keys in his 1878 patent 207,559, and for the tenacity of this layout, remain obscure.

A 1915 ergonomic keyboard

Soon after Sholes' keyboard became widely used, it surely must have become (painfully) clear that the design posed major use problems for the typist. In 1915, Heidner obtained US Patent 1,138,474. On page 1 he announced that he had "invented certain new and useful improvements" in keyboards "to enable the operator to obtain a better view of the keys and to write with greater ease, in a less cramped position than ordinarily." He states: "With this object in view, I divide the keyboard into

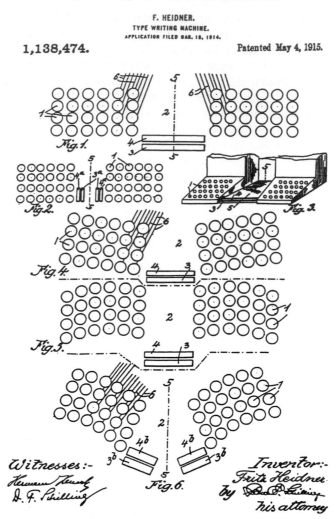

F. HEIDNER.
TYPE WRITING MACHINE.
APPLICATION FILED MAR. 18, 1914.

1,138,474.   Patented May 4, 1915.

Figure 5.9   Illustration from Heidner's 1915 patent. This seems to include about
every major ergonomic improvement of the keyboard as described in the
literature published during the remainder of the twentieth century.

halves and locate the two groups of keys thus formed in such manner that
the forearms of the operator ... instead of converging, lie substantially
parallel with each other. ... [T]he hands have not to be twisted outward
... and there being thus much less strain upon the abducent muscles,
writing is rendered considerably less fatiguing."

In addition to splitting the keyboard into left and right halves and
placing them at a slant angle to allow better forearm posture, Heidner
also arranged the keys in curved rows "in accordance with the natural
form of the hand, that is to say, lengths of the fingers. ... [C]onvergence

of the key groups further facilitates operation of the keys by the fingers in their natural position in the extended axis of the forearm." Figure 5.9 is taken from his 1915 patent and shows his design recommendations.

Heidner's astute observations preceded scientific research by at least a decade, such as published by Schroetter (1925) and Klockenberg (1926), who measured effort and recorded fatigue associated with typing. Proposals similar to Heidner's re-appeared later; examples include hypotheses in the 1920s by Klockenberg, in the 1940s by Dvorak and Griffith, in the 1960s by Kroemer and thereafter by many other inventors of improved keyboards in the last decades of the twentieth century. Figure 5.10 shows a keyboard that is split into halves for each hand. The sections can be tilted down to the sides. The keys are arranged to follow the natural motion paths of the hands' digits.

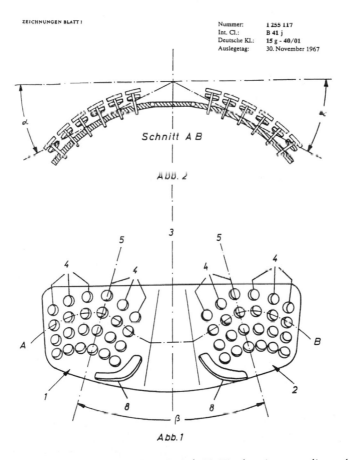

Figure 5.10 German Patent 1 255 117 (of 1967) showing a split and tiltable keyboard on which the keys are arranged to follow the motion paths of the hands' digits.

The listing at the end of Chapter 6 provides a time line of designs meant to overcome the problems associated with flat keyboards based on Sholes' invention.

Problems with current conventional keyboards

Certainly, the most serious problems with the use of keyboards, mechanical or electronic, stem from the excessive motions of the wrist and hand digits to accomplish the needed key activations. A keyboarder, typing at a speed of 50 words a minute (with one word containing 5 letters) for 6 hours a day, bends the fingers 90,000 times to press down keys, followed by the same number of finger elevations to release the keys. Sixty words for 8 hours would mean 144,000 key presses and 144,000 return movements. Even a slow keyboarder, typing only 30 words a minute, completes 36,000 digit bends and 36,000 digit lifts in just 4 hours. Many hands cannot take such frequent mechanical strain and respond with irritation and possibly inflammation.

Other problems that commonly arise are related to design details of the keyboard. They include:

- zigzag key columns on the QWERTY key pad, but straight columns on other keypads: hard on fingers and mind;
- straight rows of keys: fingertips are not similarly aligned, strain and fatigue result;
- horizontal rows of keys that enforce extreme inward rotation (pronation) of the forearms: strain and fatigue;
- large numbers of keys (more than 100, up to 130, on current full size keyboards) that require extreme digit and wrist motions and sideways stretch of the little finger: strain and fatigue;
- no support for holding arms and hands over the keyboard: strain and fatigue.

More details are provided by Kroemer et al. (2001) and are contained in the listing in Chapter 6.

Ergonomics of data entry

Designing the motor interface

An operator using current technology does most of the data transmission to the computer by hand: the common interface is the conventional flat keyboard, often accompanied by other input means such as mouse, trackball, joystick, or light pen. The design of these interfaces (discussed in some detail in several chapters in the handbook edited by Helander and Landauer, 1997) affects the workload and the motor activities of the operator, and dictates the layout of the computer workstation.

Figure 5.11   The typical mechanical typewriter, circa 1930 to 1950.

Keyboard

Unfortunately, the keyboard of the old typewriter is still used, essentially unchanged in layout but usually much enlarged, as the major input device for computers. The conventional QWERTY keyboard has several un-ergonomic features, several of which are discussed above.

Overuse disorders are common in keyboard operators. Causal or contributing factors are the frequency of key operation combined with awkward forearm and wrist postures, especially pronation and lateral deviation (Hagberg and Rempel, 1997; Keir et al., 1996; Martin et al., 1996; National Research Council, 1999; Rempel et al., 1992, 1999).

The keyboards of the old mechanical typewriters (see Figure 5.11) had strong key resistances and required large key displacements. The effort expended could overwork especially the weaker fingers, such as the little fingers. Suggested improvements include relocation of the letters on the keyboard and new geometries of the keyboard, such as curved arrangements of the keys. Other proposals divided the keyboard into halves, one for each hand, arranged so that the center sections are higher than the outsides, thus avoiding the pronation of the hand required on the flat keyboard. Another idea involves activating two or more keys simultaneously, called chording, to generate one character or whole words, or chunks of words. Such machines are routinely used by court reporters. Of course, keys do not have to be of the conventional binary (on/off) tap-down type but may be toggled or turned with three or more different contact positions. Figure 5.12 shows an example of an ergonomic keyboard, the Ternary Chord Keyboard (Langley, 1988, US Patent No. 4,775,255) which has only four keys for each hand (McMulkin and Kroemer, 1994). It allows the hand to rest on built-in wrist pads, and

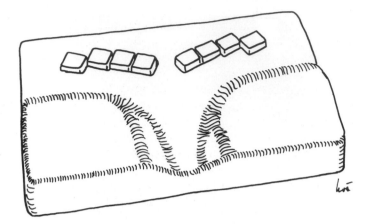

Figure 5.12   An example of an ergonomic keyboard, the Ternary Chord Keyboard, US Patent 4,775,255. It has only 8 keys, and built-in wrist rests.

the keys can support the fingers because the keys are toggled, not tapped down, as is the norm (see Figure 5.13).

Overviews of past keyboard designs have been compiled, e.g., by Kroemer (1972, 1996, 1997, 2001), Alden et al. (1972), and Noyes (1983a,b). With new technologies, sensors, and activation devices, means can be envisioned in the near future to transfer input signals to computers without using keyboards.

With the advent of notebook and laptop computers, small keyboards with much fewer than the 100+ keys of the 1990s have become commonly accepted again. (The typewriter of old had less than 50 keys.) One can hold small key pads in the hand, on the lap, in an armrest, in a glove, even

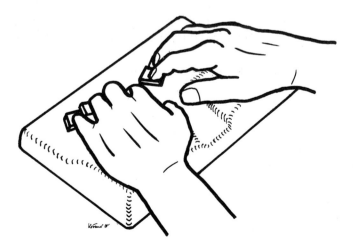

Figure 5.13   The hands can rest on the Ternary Chord Keyboard.

on the shell of a space suit. New developments may radically change the nature and appearance of our current old-fashioned keyboards and freely permit suitable arm and hand postures, in contrast to the restricted postures with the large keyboards of the late twentieth century.

New keypad designs have their opponents. Arguments brought forth against human-engineered key sets included that their posture and health advantages were unproven, that they required prolonged re-learning, and that throughput was worse, or at least not better than with the old-fashioned keyboard. However, these arguments have proven to be exaggerated or even false; one quickly adjusts to ergonomically designed keyboards, they are advantageous for body posture, and keying performance is at least equal to that on conventional keyboards (Keller et al., 2000; Marklin et al., 1997, 1999; Simoneau et al., 1999; Smith et al., 1998; Swanson et al., 1997).

Other input devices

In addition to the traditional key (discussed above), a variety of other input devices can be used, including:

- *mouse*: a palm-sized, hand-contoured block with one or more finger-operated button(s) commonly slid on a surface (mouse pad), mostly used to move a cursor;
- *puck*: similar in shape to a mouse but typically has a reticular window used on a digitizing surface (tablet);
- *trackball*: a ball mounted in an enclosure whose protruding surface is moved by palm or hand digits, usually to move a cursor;
- *joystick*: a short lever, operated by a fingertip (such as a nipple among keys of a keyboard) or, if larger, by the hand, typically for moving a cursor for pointing or tracking;
- *stylus*: a pencil-shaped, hand-held device often used for object selection, free-hand drawing, and cursor movement, usually on a tabletop digitizing surface;
- *light pen*: similar to a stylus but commonly used on a cathode ray tube (CRT) display surface;
- *tablet*: a flat, slate-like panel over which a stylus or puck is moved, usually for cursor movement and object selection;
- *overlay*: an opaque overlay of a tablet that provides graphics;
- *touch pad*: an area that translates the motion of the touching finger into a corresponding motion of the cursor;
- *touch screen (touch sensitive panel)*: an empty frame, or overlay, mounted over the display screen, that locates the position of a finger or pointing device, used for object selection, object movement, or drawing.

Typical uses of these devices are to move a cursor, input a single character or number, manipulate the screen content, digitize information, and point to insert or retrieve information; but the tasks change with evolving technology and software. Therefore, any ergonomic recommendations, including the new ANSI Standard, become outdated quickly, and the designer must follow the current literature closely.

The design of the keyboard or any other manipulated input device, especially of a mouse or puck, determines hand and arm posture. The essentially horizontal surface of conventional keyboards and of most mice requires that the palm also be kept approximately horizontal; this is uncomfortable because it requires strong pronation in the forearm, close to the anatomically possible extreme. A more convenient angle of forearm and wrist rotation is achieved by a sideways-down tilt of the mouse surface that is in contact with the palm. This is similar to splitting a keyboard and tilting the sections down to the side.

Hands and wrists are not the only body parts affected by the designs of keys and other input devices. The location of hand-operated input devices of any kind naturally determines the position of the hand and of the forearm and of the upper arm and, consequently, even affects the posture of the trunk. As a rule, the best position for the hand is in front of the body, at elbow height. If mouse, puck, stick or other input instruments are used jointly with a keyboard, they should all be placed closely together; in fact, many keyboard housings already contain a trackball, touchpad or nipple.

Supporting arm and hand in suitable fashion relieves shoulder and back muscles from their holding tasks. Various kinds of arm, wrist, and hand rests can be employed during the work, or at least during breaks in extended work. Such rest periods are helpful but by themselves are not effective in overcoming unsuitable equipment and excessive workloads.

New solutions

Traditionally, computer entries have been made by mechanical interaction between the operator's fingers and such devices as key, mouse, trackball, or light pen. These entry techniques incorporate a complex task sequence:

1. the breakdown of formulas, sentences and words into letters, numerals, and symbols;
2. the association of each letter or numeral or symbol to a distinct key, other device or function element;
3. the manual operation of keys and the like for separate entry of the numerals, symbols and letters into the computer.

The computer program then re-constructs the original words, sentences and formulas. Obviously, this is a time-consuming, complicated use of the human's mind and involves intensive manual work. In a sense, efforts of both mind and body are wasted.

It would be much better to transmit the message from the human to the computer as a whole, at least in batches. For this, voice communication with the computer is an apparent and (now) technically feasible solution.

If we think about it, it becomes readily apparent that there are many other means to generate input to the computer. Consider, for example, the use of

- hands and fingers for pointing, gestures, sign language, tapping;
- arms for gestures, making signs, moving or pressing control devices;
- feet for motions and gestures, for moving and pressing devices;
- legs for gestures, moving and pressing devices;
- torso, including the shoulders, for positioning and pressing;
- head, also for positioning and pressing;
- mouth for lip movements, use of the tongue, or breathing such as through a blow/suck tube;
- face for grimaces and other facial expressions;
- eyes for tracking.

Combinations and interactions of these different input signals could conceivably be used in a similar way in which we utilize them during face-to-face communication. Of course, the techniques developed to recognize these signals must be able to clearly distinguish them from environmental clutter or other loose energy that could interfere with sensor pickup.

Such ideas are not far-fetched, either in sensor technology or in human input activities (Greenstein, 1997). Even such simple measures as returning to a small set of keys, as on the old QWERTY, or implementing key chording together with voice commands, all while resting the wrists on suitable pads during frequent breaks, might make computer input much more efficient and, at the same time, alleviate many of the operators' cumulative trauma disorders discussed earlier.

ANSI/HFS 100 (1988, with its new edition expected in 2001) refers to specific keyboard design features such as key spacing, key size, key actuation force, and key displacement. It also points to overall design dimensions for mouse and puck input devices, trackballs, joysticks, styli and light pens, tablets and overlays, and touch sensitive panels. Keep in mind, however, that the ANSI standard merely specifies acceptable applications based on accepted human factors engineering research and experience and that the standard does not apply to operator health considerations or work practices. The standard permits alternate computer workstation technologies and will not impede development and use of novel solutions.

Changes come with technical developments

Some of the current computer work tasks probably should not be imposed on humans, such as the prolonged simple input of numbers or of word texts. Not only is this boring, but repetitive finger movements may be a source of cumulative trauma disorders, such as tendonitis in hands or arms or carpal tunnel syndrome (see above and Chapter 6). Repetitive entry should be automated, e.g., by machine recognition of characters, by scanning, or by voice recognition. Feedback from the computer about its memory content can also be given to the operator through acoustical or sensory means other than just by display on the screen.

Clever software offers possibilities to facilitate the task of the computer operator, such as automated programs for grammar and spelling, stringing of characters, and algorithms that check and indicate outliers in data or unusual events or repetitive occurrences. Certainly, a wide variety of opportunities exist to change and, one hopes, to improve and facilitate the work of the computer operator.

As discussed in Chapter 2, many people in the office feel proud to have autonomy in performing their work, to take responsibility for its quality and quantity, and to control their timing. Most prefer to perform larger tasks from beginning to end instead of simply doing specialized tidbits. Within the limitations set by the need to have certain work done at a given time, the employee should be free to distribute the workload, both in amount and in pace, according to his or her own preferences and needs.

Organizing working time, particularly providing changes in work periods and rest pauses, is important for many computer tasks. Most people are bored by repetitive, monotonous, and continuous work; instead, varying tasks of different lengths should be provided. In this way, the computer operator has occasion and cause to shift from one task to another, to move away from the computer for a while, and to simply do something else, or to take a break. In other words: providing a new input device without considering associated factors of the work does not in itself create a better office (see also the first two chapters of this book). Instead, organizing the work differently, keeping in mind the needs and preferences of the operator, and ensuring that the physical set-up of a new device is appropriate and comfortable, are all critical.

ERGONOMIC RECOMMENDATIONS

Ergonomic keyboards

Consider using a modern ergonomic keyboard, for example; one that

- has a split and slanted QWERTY (alpha) key section;
- has a split QWERTY section whose halves are tilted down to the sides;

- has a key feel that you like (mostly in terms of key resistance and travel);
- has built-in cursor controls (such as track pad, ball, nipple) that you find easy to operate;
- is designed and built to the newest technology and human engineering standards (such as ANSI/HFES 100);
- has only as many keys as you need.

Conventional keyboards

If you continue to use a conventional keyboard, select one that

- has a key feel that you like (mostly in terms of key resistance and travel);
- has built-in cursor controls (such as track pad, ball, nipple) that you find easy to operate;
- is designed and built to the newest technology and human engineering standards (such as ANSI/HFES 100);
- has only as many keys as you need.

Special key sets

If you need a special key set occasionally, such as a numeric pad, consider getting one in addition to your regular keyboard and place that special keypad carefully in front of you; do not compromise between them.

Try out as many keyboards as you can. To really evaluate them you may have to use them for hours or even days. Do not hesitate to discard a keyboard if you do not like it; keyboards cost much less than treating repetitive strains does. Consider using a laptop with its smaller keyboard instead of a desktop computer.

Other input devices

Consider voice communication with your computer. Evaluate carefully whether to use

- mouse or trackball or trackpad or nipple for cursor control;
- puck or joystick or stylus or light pen or tablet or touch screen for pointing, tracking, digitizing, drawing and other special tasks.

This feels good

Once you have selected the input device that you really like (see above), sit in a truly comfortable chair (see Chapter 4) and place the input device

- directly in front of you (with the display closely behind it);
- at elbow height (with your upper arms hanging from your relaxed shoulders);
- over your thighs.

You may consider working standing up, at least once in a while. For this you need an easily adjustable workstation, or a second computer set up for standing operation. A laptop or handheld computer allows much variety.

Use a padded wrist rest (or armrest) at least during inputting breaks while you remain at your workstation.

Take short breaks from keyboarding or other manual inputting often. It is best to leave your workstation completely for a while.

What to do if you are not comfortable?

If your hand/wrist/arm/shoulder or eye hurt: check whether or not you followed the ergonomic recommendations listed above. If not, do. If the pain continues, or worsens, talk with your supervisor; you may need a medical evaluation.

If the hand (wrist/arm/shoulder) with which you operate the mouse (or other input device) is tired or hurts: Change to the other hand. Even if this feels awkward at first, you will soon become proficient. Then make it a rule to switch every half hour or so. If the pain continues, or worsens, talk with your supervisor; you may need a medical evaluation.

If your back/neck hurts: check whether or not you followed the ergonomic recommendations listed above, especially regarding seat and placement of input devices and display (check also the recommendations in Chapters 4 and 8). If not, do. If the pain continues, or worsens, talk with your supervisor; you may need a medical evaluation.

If you feel tired or bored: Take breaks (or get another job).

No pain, no gain? Coach was wrong: Computer and other work should not cause you pain; not in the hands, arms, shoulders, neck, back, or eyes. Get rid of conditions that cause discomfort; it gains you nothing but is likely to harm your work output and, worse, your health.

References

Adler, M. (1997). *Antique Typewriters. From Creed to QWERTY*. Atglen, PA: Schiffer.

Alden, D. G., Daniels, R. W. and Kanarick, A. F. (1972). Keyboard Design and Operation: A Review of the Major Issues. *Human Factors 14*, 275–293.

ANSI/HFS 100 (1988). *American National Standard for Human Factors Engineering of Visual Display Terminal Workstations*. Santa Monica, CA: Human Factors Society.

ANSI/HFES 100 (2001). *US National Standard for Human Factors Engineering of Computer Workstations*. Santa Monica, CA: Human Factors and Ergonomics Society.

Armstrong, T. J. and Lackey, E.A. (1994). *An Ergonomics Guide to Cumulative Trauma Disorders of the Hand and Wrist*. Fairfax, VA: American Industrial Hygiene Association.

Greenstein, J. S. (1997). Pointing Devices, in M. Helander, T. K. Landauer and P. Prabhu (Eds.) *Handbook of Human-Computer Interaction* (2nd ed.). Amsterdam: Elsevier, 1317–1348.

Hagberg, M. and Rempel, D. (1997). Work-Related Disorders and the Operation of Computer VDTs, in M. G. Helander, T. K. Landauer and P. V. Prabhu (Eds.), *Handbook of Human-Computer Interaction* (2nd ed.). Amsterdam: Elsevier, 1415–1429.

Heidner, F. (1915). *Type-Writing Machine*. Letter's Patent 1,138,474, dated May 4, 1915; application filed March 18, 1914. United States Patent Office.

Helander, M. G., Landauer, T. K. and. Prabhu, P. V (Eds.) (1997). *Handbook of Human-Computer Interaction* (2nd ed.). Amsterdam: Elsevier.

Herkimer County Historical Society (Ed.) (1923) The Story of the Typewriter 1873–1923. Published in Commemoration of the Fiftieth Anniversary of the Invention of the Writing Machine. Herkimer, NY: Herkimer County Historical Society.

Keir, P. J., Bach, J. M., Engstrom, J. W. and Rempel, D. M. (1996). Carpal Tunnel Pressure: Effects of Wrist Flexion/Extension, in *Proceedings, American Society of Biomechanics 20th Annual Meeting*. Atlanta, GA: American Society of Biomechanics, 169–170.

Keller, E., Fleischer, R. and Strasser, H. (2000). Subjective Assessment of the Ergonomic Quality of a Keyboard for VDU Workplaces, in *Proceedings of the XIVth Triennial Congress of the International Ergonomics Association and 44th Annual Meeting of the Human Factors and Ergonomics Society*. Santa Monica, CA: Human Factors and Ergonomics Society, 6344–6347.

Klockenberg, E. A. (1926). *Rationalization of the Typewriter and of Its Use* (in German). Berlin: Springer.

Kroemer, K. H. E. (1972). Human Engineering the Keyboard. *Human Factors 14*, 51–63.

Kroemer, K. H. E. (1992). Avoiding Cumulative Trauma Disorders in Shop and Office. *American Industrial Hygiene Association Journal 53*, 596–604.

Kroemer, K. H. E. (1996). Manual Data Entry: Yesterday, Today, Tomorrow? in *Proceedings, ErgoCon '96, The Silicon Valley Ergonomics Conference and Exposition*. San Jose, CA: San Jose State University, 3–4.

Kroemer, K. H. E. (1997). Design of the Computer Workstation, in M. G. Helander, T. K. Landauer and P. V. Prabhu (Eds.), *Handbook of Human-Computer Interaction* (2nd ed.). Amsterdam: Elsevier, 1395–1414.

Kroemer, K. H. E. (2001). Keyboards and keying. An annotated bibliography of the literature from 1878 to 1999. *International Journal Universal Access in the Information Society*. Berlin: Springer, in press.

Kroemer, K. H. E., Kroemer, H. B. and Kroemer-Elbert, K. E. (2001). Ergonomics: How to Design for Ease and Efficiency (2nd ed.). Upper Saddle River, NJ: Prentice Hall.

Kuorinka, I. and Forcier, L. (Eds.) (1995). *Work Related Musculoskeletal Disorders: A Reference Book for Prevention*. London Taylor & Francis.

Langley, L. W. (1988). *Ternary Chord-Type Keyboard*. Patent No. 4,775,255. United States Patent Office.

Marklin, R. W., Simoneau, G. G. and Monroe, J. F. (1997). The Effect of Split and Vertically-Inclined Computer Keyboards on Wrist and Forearm Posture, in *Proceedings of the Human*

Factors and Ergonomics Society 41st Annual Meeting. Santa Monica, CA: Human Factors and Ergonomics Society, 642–646.

Marklin, R. W., Simoneau, G. G. and Monroe, J. F. (1999). Wrist and Forearm Posture from Typing on Split and Vertically-Inclined Computer Keyboards. *Human Factors 41*, 559–569.

Martin, E. (1949). *Die Schreibmaschine und ihre Entwicklungsgeschichte (The Typewriter and its Development History)*. Aachen, Germany: Basten.

Martin, B., J. Armstrong, T. J., Foulke, J. A., Natarajan, S., Klinenberg, E., Serina, E. and Rempel, D. (1996). Keyboard Reaction Force and Finger Flexor Electromyograms During Computer Keyboard Work. *Human Factors 38*, 654–664.

Max-Planck Gesellschaft (1967). Patentschrift. (West) German Patent 1 255 117.

McMulkin, M. L. and Kroemer, K. H. E. (1994). Usability of a One-Hand Ternary Chord Keyboard. *Applied Ergonomics 25*, 177–181.

Moon, S.D. and Sauter, S.L. (Eds.) (1996). *Beyond Biomechanics: Psychosocial Aspects of Musculoskeletal Disorders in Office Work*. London: Taylor & Francis.

National Research Council (Ed.) (1999). *Work-Related Musculoskeletal Disorders: Report, Workshop Summary, and Workshop Papers*. Washington, DC: National Academy Press.

Noyes, J. (1983a). The Qwerty Keyboard: A Review. *International Journal of Man-Machine Studies 18*, 265–281.

Noyes, J. (1983b). Chord Keyboards. *Applied Ergonomics 14*, 55–59.

Putz-Anderson, V. (1988). *Cumulative Trauma Disorders: A Manual for Musculoskeletal Diseases of the Upper Limbs*. London: Taylor & Francis.

Rempel, D. M., Harrison, R. J. and Barnhart, D. (1992). Work-Related Cumulative Trauma Disorders of the Upper Extremity. *Journal of the American Medical Association 267*(6), 838–842.

Rempel, D., Tittiranonda, P. Burastero, S., Hudes, M. and So, Y. (1999). Effect of Keyboard Keyswitch Design on Hand Pain. *Journal of Occupational and Environmental Medicine 41*, 111–119.

Schroetter, H. (1925). Knowledge of the Energy Consumption of Typewriting. *Pflueger's Archiv Gesamte Physiologsche Menschen Tiere 207*(4), 323–342 (in German).

Sholes, C.L (1878). *Improvement in Type-Writing Machines*. Letter's Patent No. 207, 559, dated August 27, 1878; application filed March 8, 1878. United States Patent Office.

Simoneau, G. G., Marklin, R. W. and Monroe, J. F. (1999). Wrist and Forearm Postures of Users of Conventional Computer Keyboards. *Human Factors 41*, 413–424.

Smith, M. J., Karsh, B. T., Conway, F. T., Cohen, W. J., James, C. A., Morgan, J. J., Sanders, K. and Zehel, D. J. (1998). Effects of Split Keyboard Design and Wrist Rest on Performance, Posture, and Comfort. *Human Factors 40*, 324–336.

Swanson, N. G., Galinsky, T. L., Cole, L. L., Pan, C. S. and Sauter, S. L. (1997). The Impact of Keyboard Design on Comfort and Productivity in a Text-Entry Task. *Applied Ergonomics 28*, 9–16.

Treaster, D. E. and Marras, W. S. (2000). A Biomechanical Assessment of Alternate Keyboards Using Tendon Travel, in *Proceedings of the XIVth Triennual Congress of the International Ergonomics Association and 44th Annual Meeting of the Human Factors and Ergonomics Society*. Santa Monica, CA: Human Factors and Ergonomics Society, 6685–6688.

6   Feeling good at work

Overview

Doing well in the office is not solely related to the major activities, such as keyboarding. Other tasks are likely to take a lot of effort and time, even if you discount them as not so important. And performing well depends on more than employee skills and corporate reward incentives. Instead, we must also feel safe and secure in the office, with work conditions that do not expose us to dangers and hazards. You probably do not routinely ask yourself these questions when you step into your office in the morning, but is your company doing whatever it can to keep your work environment free of unnecessary hazards? Are you concerned about an outbreak of fire? Do cables and electrical cords get in your way? Do you run all over the place to retrieve old files? How much do you use the telephone? Does your neck hurt, or your wrist?

Fire safety

Fire is a threat to the people in the office, their personal belongings and their work products, as well as to equipment and the building. Fire safety is therefore a high-priority demand. Fire produces not only heat but also toxic fumes and smoke, all dangerous threats.

The avoidance of fire hazards is a special technology, but alerting people in the office to an existing combustion is an ergonomic challenge. The first task is to alert people to the existence of a danger, the second task to guide them away from it.

As the specialized literature describes in much detail, a combination of acoustic and visual alarm signals works most reliably to inform people of danger. Usually, a loud (high intensity) signal is used that pierces the regular noise level through its carefully selected sounds (in terms of frequencies), modulated over time. The sound signal is best accompanied by flashing red lights. Typical everyday examples are the sirens and strobe lights used on police and firefighting vehicles.

Escape paths must be kept open and clearly marked, with the indicators clearly visible even in smoke, in crowded conditions, and in panic situations. Emergency lighting is necessary. The escape routes must be usable by disabled persons, including by those in wheelchairs. Fire drills are a must, even if they appear to be a nuisance.

Security

Personal security, and the security of our belongings from theft, can be of great concern to those who work in offices that are open to the public, especially if they are often alone, work during unusual hours or work in unsafe neighborhoods. Formal surveillance and guarding systems at the perimeter of the office may be necessary, with limited and supervised entry and exit doors (and often including outside walkways and parking lots). Within the office, informal arrangements can be helpful, including avoiding needless high visual barriers and planning entrance and exit routes for visitors that are easily viewed by the office occupant. Of course, a reliable alarm system that brings help quickly is essential.

Theft rather than personal attack is a problem that is all too common. The surveillance and guarding arrangement just mentioned is effective here as well, but safe places to lock personal and company property also help to assure feelings of ease and security at work.

Cabling and wiring

The need for data transmission, telephone lines, and electricity outlets per workstation has increased tremendously with the magnified utilization of electronic communication. It is now common to see two displays, a printer, a scanner plus the usual telephone at a regular office workstation – and there is no way to foresee the changes that are likely to come with more advanced communications and developing technologies. (Is it hoping for too much that all data and energy transmission become wireless?)

The cables and cords and wires can come to the workstation through a hollow dropped ceiling and power poles, via hollow accessible floors, through pre-set ducts in floor and wall, or as flat bundles under carpet (with the danger of tripping). At the workstation, cables and wires can run inside stationary furniture or along its side – but at some point, they come into the open to connect to the equipment.

While the provision of cables, cords, and wires seems to be mainly a task for management, architect, and builder, it affects the employee in terms of safety, intrusion into the workspace (no stumbling, please), orderliness and cleanliness of the workplace, and the ability to perform tasks. It affects the general appearance of the workplace and hence one's job attitude.

Filing and keeping paper records

The 1980s promise of the paperless office via electronics was premature, unfortunately: with the ease of electronic printing, paper records are apparently even more voluminous today then a few decades ago. This has led to the need for storage spaces for paper files – often messy and always flammable.

Some of the records that we want to keep can be maintained on our own computers, and only rather few written materials need to be kept in, or next to, our personal workstation. Hand-held, palm-sized computers are likely to soon take over the functions of appointment calendars, of notebooks, of telephones and other communication devices.

Seen through the eyes of the ergonomist, one significant benefit can be gained from the vast needs for paper storage. Filing cabinets for the whole office should be located away from the individual workstations so that each office worker is forced to get up, walk over, look for and grasp the file needed. This serves as means to make people stop sitting and, instead, to move their bodies: exercise by office layout. So there is indeed something good about office paperwork.

Communicating

The ways in which we communicate with others have changed, and will undoubtedly continue to change. In just a few decades, the telephone made the telegraph obsolete, pneumatic mail delivery within a building is gone, telegrams and the telegram-style verbiage are nearly forgotten, and the amount of mail delivered by the postal services to addresses all over the globe has been reduced by fax and e-mail. New ways to communicate change the tasks and talents of office personnel; there are hardly any secretaries left who take dictation via shorthand; instead, they and their bosses learned how to do word processing. Further development in electronic information storage and transmission will lead to more demands for new facilities and skills.

The telephone is one old office tool that is still in wide use. It has become one piece, acquired push buttons, gone wireless, and its quality of transmission has much improved. But it is still difficult to hold against the ear and close to the mouth, as Figure 6.1 illustrates. This is particularly true when we talk while keyboarding or using our hands for another manual task – and, in the process, we cradle our phone between shoulder and ear, craning our neck and twisting our shoulder to keep the receiver snuggly in place. Using a device that is carried on the head, probably with a separate mike close to the mouth and the speaker at one ear, avoids the twisting of the upper body, frees both hands, and makes for clearer communication.

Figure 6.1   The old-fashioned telephone – a pain in the shoulder and neck. (Illustration courtesy of Herman Miller, Inc.)

Looks and appearances

Esthetics are important. A well laid-out office is appealing, whether it is a large room or a small cubicle (see Chapter 3). Even a hardboiled grump prefers a beautifully furnished, appealingly colorful, well-lit, clean and orderly office to an ugly, dark, dirty and cluttered one. Most persons like to give their workplace a personal touch by re-arranging fixtures and furniture to their preferences, adding colors and plants, placing pictures and photos, bringing in items that they cherish. This helps to feel well at work and feel good about work, and – directly or indirectly – to work with ease and efficiently.

Avoid musculoskeletal diseases related to keyboarding

Keyboarding, whether done on a typewriter, calculator, desktop or laptop or notebook computer, can lead to various kinds of body discomfort – often in the hands and wrists, or in the shoulders. If the feeling of strain goes away during a break, then you may just want to check whether you could do your keyboarding more correctly; look through Chapter 5. But if your hands, arms and shoulders continue to be bothersome, then talk with your supervisor about help and advice for better work assignments and workplace arrangements. If persistent tingling, numbness, or pain develop, see an orthopedist.

You must avoid serious overexertion of your musculoskeletal body apparatus. An overuse disorder can show slowly over days or weeks or months, or it may appear fairly suddenly, even if you like the activity that you are overdoing, such as playing the piano, or even swinging your golf club. First signs of overuse are slight pains and occasional aches; in the hand you may experience numbness and tingling. Continuous pain then may ensue, often with symptoms of inflammation and swelling, especially of tendons and their sheaths. Cumulative trauma disorders (CTDs), also often called repetitive strain injury (RSI), include sets of long-known health problems – see the listing below. They can be averted by use of suitable equipment and, most importantly, by not overworking the body. When you do experience the initial signs of an overuse injury, stop doing what is causing the problem, talk with your supervisor, and get medical treatment as needed.

Occupational diseases

Three hundred years ago, Bernardino Ramazzini, the "Father of Industrial Hygiene", wrote about occupational injuries in offices (see the translation of the Latin text by Wright, 1993). Of course, in the early 1700s, there were no typewriter or computer keyboards yet, but muscle pain and cramps were occurring among secretaries and scribes, stemming from repetitive tasks that involved long-maintained hand and arm postures. Ramazzini also described many other occupational diseases, and in the mid-1800s, repetitive strain disorders were well known to be correlated with certain occupations, for example in textile workers (as stated by Burry and Stoke, 1985) or among musicians, especially pianists (as described by Fry, 1986a–d; Hochberg et al., 1983; Lockwood, 1989). In the 1860s and 1870s, Sholes and his co-inventors obtained several US patents for typewriters with keyboards derived from that of the piano. Sholes' last patent (No. 207,559 of 1978) had four straight rows of eleven round key buttons each (see Chapter 5 for more information). The third row (as counted from the operator) started, from the left, with the letters QWERTY inscribed on the keys – an arrangement still used on today's computer keyboards for the English language, and probably one of the reasons for repetitive injuries to keyboarders (Kroemer, 2001).

In the 1890s, a variety of occupation-related repetitive diseases were described in the literature. The factors that cause, aggravate or precipitate CTDs can be part of occupational or leisure activities, as indicated by such names as

- writer's cramp or scribe's palsy
- goal-keeper's, seamstress' or tailor's finger
- bowler's, gamekeeper's or jeweller's thumb
- bricklayer's hand

- meatcutter's, musician's, pianist's, stitcher's, tobacco-primer's, telegraphist's or washerwoman's wrist
- carpenter's arm
- carpenter's, jailer's or student's elbow
- porter's neck
- shoveler's hip
- weaver's bottom
- housemaid's or nun's knee
- ballet dancer's or nurse's foot.

In the twentieth century, this list was extended with new descriptive terms:

- baseball-catcher's hand
- typist's, cashier's, letter sorter's and yoga wrist
- golfer's, tennis and mouse elbow
- letter carrier's shoulder
- carpet-layer's knee.

Cumulative trauma disorders

The prominent reasons for such diseases are highly repetitive activities, often with awkward positions and movements of the body segments involved, and pressure from equipment on the body – like that often associated with keyboarding. A single such small trauma is not injurious to the body if it occurs occasionally, but the cumulative effects of microtrauma can lead to overexertions.

Here is a formal definition: cumulative trauma disorders (CTDs) are regional impairments of muscles, tendons, tendon sheaths, ligaments, nerves, and joints associated with activity-related repetitive mechanical trauma.

Other names used are repetitive trauma injury (or illness, or disorder), repetitive motion injury (or illness, or disorder), repetitive strain injury (RSI), (occupational) overuse disorder (OD), work-related musculoskeletal disorder (WRMD or WRMSD).

Medically, CTDs are usually classified as irritation, sprain, strain, or inflammation. Terms used to describe their nature include tendinitis (tendonitis), tenosynovitis, myalgia, bursitis, peritendinitis, epicondilitis, brachial plexus syndrome, thoracic outlet syndrome, carpal tunnel syndrome (CTS), regional pain syndrome, neurovascular syndrome, cervicobrachial syndrome, compression syndrome, entrapment syndrome, and neuropathy.

What was known about CTDS and when?

As Ayoub and Wittels (1989), Burnette and Ayoub (1989), Burry and

Stoke (1985), Fry (1986a–d), Hochberg et al. (1983), and Lockwood (1989) described in detail (Table 6.1), activity-related causes of CTDs were extensively reported in the contemporary literature, and their physiological-pathological mechanisms were extensively discussed by the end of the nineteenth century.

Further insights in the pathology of cumulative trauma disorders were gained in the first four decades of the twentieth century. CTDs were now firmly linked to physical strain of the musculoskeletal system, with the stress generated not so much by the amount of energy expended in the single muscular activations but rather by the accumulation of efforts in highly repetitive work, as explained by Schroetter (1925), Klockenberg (1926), Conn (1931, 1934) and especially by Hammer (1934). Cumulative trauma disorders were understood as related to design features and operation of equipment, such as hand tools, telegraphs, and the typewriter.

The work required of the typists' hands and arms to pounce on the keyboard of mechanical typewriters was fairly heavy, given the large displacement of the keys and their stiff resistance. Furthermore, the arrangement of the keyboard itself, and of the keys on it, still followed Sholes' original design, which was certainly not ergonomic because it

Table 6.1 Early publications related to cumulative trauma disorders, keying and keyboard design (Kroemer, 2001)

Year	Author(s)	CTD aspects	Keying and keyboard design
1713		Ramazzini (transl. Wright 1993)	Diseases of workers due to "violent and irregular motions and unnatural postures of the body"
1872	Poore	Discussion of the muscular causes for writers' cramp	Cramps common in pianists and telegraph operators
1868–78	Sholes		Various keyboard designs shown in patents, including one – the last – with QWERTY layout
1887	Poore	Discussion of 21 cases of muscular syndromes in pianists	Overuse due to keying
1892	Osler	Continuous and excessive use of muscles in movements followed by spasm or cramp	Spasms and cramps common in pianists and telegraphists

Table 6.2   Publications from 1900 to 1940 related to CTDs, keyboarding and keyboard design (Kroemer, 2001)

Year	Author(s)	CTD aspects	Keying and keyboard design
1909	Rowell		Proposal for new keyboard design to overcome known disadvantages
1915	Heidner		Patent of a split keyboard design to overcome motion and posture problems
1920	Banaji		Patent of a new key layout to overcome known disadvantages
1920	Nelson		Patent of a new key arrangement to overcome known disadvantages
1920	Wolcott		Patent of a new key layout design to overcome known disadvantages
1924	Hoke		Patent of a new split keyboard design to overcome motion problems
1924	Book		High motor ability of upper extremity necessary for champion typists
1925	Schroetter	Measurements of heart rate and oxygen consumption while typing	
1926	Klockenberg	Repetitive key operation and posture cause serious problems in arms and hands	Unhealthy posture while keying, due to keyboard design. Proposes new design to overcome problems
1927	Zollinger	Seven cases of tendovaginitis after repetitive actions (929 cases of tendovaginitis mentioned)	

Table 6.2 (*continued*)

Year	Author(s)	CTD aspects	Keying and keyboard design
1930	Gilbert	Finger movements among keys are difficult	Proposal for a new keyboard design to overcome motion problems
1930	Marloth		Patent of a split keyboard with keys on each in rows perpendicular to natural positions of the forearms to achieve health and comfort for the typist
1931	Conn	Repetition-related tenosynovitis is an occupational disease subject to compensation	
1934	Biegel		Patent of a new keyboard design to avoid finger overloads
1934	Hammer	High-speed hand operations predispose for tenosynovitis	
1936	Dvorak		Patent of a new keyboard design to avoid finger overloads

forced the arms into strong inward twist (pronation) and the hands into lateral bend (ulnar deviation) and required complex motions between the ill-located keys. As Table 6.2 shows, design deficiencies were soon recognized and remedies were proposed as early as 1915 by Heidner, who split the keyboard, tilted the halves down sideways, and re-arranged the key layout. In 1920, Banaji; Nelson, and Wolcott all acquired patents for their re-arrangements of the keys to alleviate the complexity of finger movements, with Dvorak's 1936 patent the best known nowadays.

As Table 6.3 shows, by 1960, physiological/ biomechanical knowledge clearly related repetitive activities, particularly typing, to strains in the tendons and sheaths of the hand, wrist, and arm. Lundervold (1951, 1958) demonstrated that electromyography was a useful measure of muscle efforts. Pfeffer et al. (1988) state that, by 1960, carpal tunnel syndrome was the most frequently diagnosed, best understood, most easily treated entrapment neuropathy.

Table 6.3   Publications from 1941 to 1960 related to CTDs, keyboarding and keyboard design (Kroemer, 2001)

Year	Author(s)	CTD aspects	Keying and keyboard design
1942	Flowerdew and Bode	16 cases of tenosynovitis related to manual work	
1943	Dvorak	Mental tension and fatigue in typists with current keyboard design	New keyboard design to avoid finger overloads
1949	Griffith		New keyboard to achieve easier finger motions
1950	Hesh		Patent with fewer keys on the keyboard, movable keys
1951	Lundervold	Occupational myalgia in typists. EMG measurements of involved muscles	
1954	Scales and Chapanis		Effects of keyboard slope on keying performance and subjective acceptance
1955	Hagen and Peters	Tenosynovitis as a result of repeated fast motions on the job; most frequently observed in stenotypists	
1956	Strong		Testing of Dvorak versus standard keyboard
1957	Buckup	Tenosynovitis as a result of repeated motions associated with use of typewriters and other office machines	
1958	Klemmer		Development and use of a ten-key typewriter keyboard
1958	Lundervold	Muscle activities when typing recorded by EMG	

Table 6.3 (*continued*)

Year	Author(s)	CTD aspects	Keying and keyboard design
1958	Hettinger	Test of disposition for tenosynovitis	
1959	Tanzer	Carpal tunnel syndrome related to manual activities	CTS occurs in very busy secretarial work
1959	Lockhead and Klemmer		Testing of an eight-key word-writing keyboard
1960	Deininger		New key design features, force-displacement characteristics
1960	Creamer and Trumbo		Lateral declinations of the keyboard less fatiguing than horizontal arrangement

By 1980, CTDs were generally understood as resulting from a series of microtraumas which individually would not do harm. But in their cumulative effects, the microtrauma can injure musculoskeletal, vascular, and nervous systems of the human body. Founded on the knowledge of human CTD pathology, the avoidance of repetitive motions in awkward hand/arm postures, such as in typing, had become a goal of industrial hygiene. Based on that medical understanding, on practical experiences and on specific research, many well-considered recommendations were made for new designs of keys and keyboards and for their use on the typewriter (Table 6.4). The replacement of mechanical levers between key top and the strike bar with electro-mechanical devices around 1970 – and then with electronic switches and circuitry – facilitated new designs of keys and key sets. Key travel and key resistance were much reduced from those on the old typewriter. Yet, the number of keys on all-purpose electronic keyboards was now typically just over 100, more than double as many as on Sholes' 1878 design.

During the 1980s, CTDs in keyboarders, especially carpal tunnel syndrome, were becoming a widespread phenomenon. "Epidemics" of CTDs occurred in Japan and Australia, and pockets of the occupational disease were found in certain industries and services in the USA.

On conventional computer keyboards, the layout of the QWERTY section remained essentially the same as during the last decades, and

Table 6.4 Publications from 1961 to 1980 related to CTDs, keyboarding and keyboard design (Kroemer, 2001)

Year	Author(s)	CTD aspects	Keying and keyboard design
1961	Droege and Hill		Performance on electric and on mechanical typewriters
1961	Peres	Repetitive activities in hand and wrist result in strain, pain, cumulative injuries	
1961	Ratz and Ritchie		In keyboarding, motor system constraints predominate over choice reaction time
1962	Diehl and Seibel		Removal of visual and particularly auditory feedback reduces typing performance
1962	Klemmer and Lockhead		Between 56,000 and 83,000 keystrokes per day performed by IBM machine operators
1962	Seibel		On chord keyboards, increased performance is achieved by decreased response times
1962	Seibel and Rochester		Keys rearranged to reduce finger movements and effort
1962	Yllo		Right-hand keyboard was rearranged to avoid unnatural wrist and elbow angle; lowering keyboard, tilting it down to the right side reduced fatigue as determined by EMG
1962	IBM		Internal proposal for new keyboard design to overcome known disadvantages
1963	Robbins	Carpal tunnel syndrome related to wrist flexion and extension	

Table 6.4 (*continued*)

Year	Author(s)	CTD aspects	Keying and keyboard design
1963	Cornog et al.		Chord keyboard feasible for address encoding
1963	Seibel		Response times increase with number of alternatives. Reaction time on keys increases with information size
1963	Morgan et al.		Recommendations for selection and design of controls
1964a,b	Kroemer (2 publications)	CTDs related to force, displacement of keys, frequency, wrist posture while keying	New keyboard design: Split keyboard, tilt each half laterally, rearrange keys
1964	Seibel		Final keying performance cannot be predicted, but must be observed
1964	Fox and Stansfield		Keyboard design determines keying rate
1965	Bowen and Guinness		Keyboards should be designed rationally and in keeping with good human engineering practice; on such keyboards, performance differences are small
1965	Conrad and Longman		Keying feedback important while learning
1965	Galitz		Keying force, travel, activation, slope, feedback
1965	Kroemer	Typists terminated keying because of aches and complaints in arms, wrists, and fingers	Proposes tilted instead of horizontal keyboards; tilt postponed fatigue
1965	Mayzner and Tresselt		Single-letter and digram frequencies

Table 6.4 (*continued*)

Year	Author(s)	CTD aspects	Keying and keyboard design
1966a	Galitz	Fatigue associated with keying	Key force, displacement, slope reviewed
1966b	Galitz		Key force and activation feedback on Microswitch keys are important
1966	Phalen	754 cases of carpal tunnel syndrome related to finger flexion with wrist flexion; References to numerous reports	
1966	Hymovich and Lindholm	62 cases of CTD, more in females than males, with repetitive work	
1966	Tichauer	"...modern industrial health care must consider...impairment caused insidiously...by gradual, cumulative, and often imperceptible overstrain of minute body elements"	Attention must be paid to the hand-tool interface
1967	Max-Planck Gesellschaft	Muscular fatigue of keyboard operators can be reduced by proper design	Patent of a split and tilted keyboard with hand-configured key arrangements
1967	Lewin		Pointing pen/electronic keyboard as input device
1967	Smith		Mistakes made in data entry on key sets
1967	West		Deprivation of visual feedback while typing increases errors
1967	Ulich	Use of EMG to assess muscular activities	

Table 6.4 (*continued*)

Year	Author(s)	CTD aspects	Keying and keyboard design
1968	Lhose		Alphanumeric keyboard standard for U.S.A.
1968	Adams		Suitable feedback is important for key operation
1968	Keele		Theories of movement control and the importance of feedback
1969	Kinkead and Gonzales		Effects of key designs, including tactile feedback
1969	Remington and Rogers		Listing of more than 300 publications on keyboards and keying collected by IBM
1969	West		Fatigue and performance during 30-min typing
1970	Caldwell	Muscle fatigue after repeated exertions	
1970	Galitz and Laska		Manual activities of computer operators
1970	Hirsch		Standard versus alphabetic keyboard
1971	Garrett		Hand dimensions and biomechanical characteristics
1971	Klemmer		Summary of 35 articles regarding keyboard design, feedback, key force and displacement
1971	Komoike and Horiguchi	233 fatigue cases in female office workers: effects of high speed and paced rhythm; localized fatigue and pain	

Table 6.4 (*continued*)

Year	Author(s)	CTD aspects	Keying and keyboard design
1971	Phillips and Kincaid		Better key arrangement is needed and technically easy
1971	Michaels		QWERTY keyboard allows better performance than alphabetic keyboards
1971	Smith and Goodwin		Alphabetic data entry by telephone key arrangement possible
1972	Gallaway		Recommendations for key force and displacement
1972	Kroemer	Effects of key operation on operator strain and performance	Keyboard arrangement, tilt to the side investigated
1972	Phalen	598 cases of CTD	
1972	Samuel		New key designs
1972	Seibel		Key force, displacement, feedback reviewed
1972	Smith		Alternative function keys
1972	Alden et al.	Hand musculature limits keying; fatigue, key force and displacement	Review of, and recommendations for, keyboard design, key force and displacement, feedback
1973	Tichauer	Cites Ramazzini's 1713 statement about diseases due to violent and irregular motions	
1973	Yoder et al.	Muscle fatigue and wrist problems in assembly, assessment by EMG	

Table 6.4 (*continued*)

Year	Author(s)	CTD aspects	Keying and keyboard design
1974	Duncan and Ferguson	Occupational cramps and myalgia associated with adverse postures of keyboarders; 90 patients with cramps and myalgia in keyboard operation; effects of keyboard design and arrangement on posture	New key and keyboard arrangement
1974	Ferguson and Duncan	Adverse postures of digits, wrists, arms, shoulders, neck, trunk due to keyboard layout and arrangement	Key arrangement, division of keyboards for each hand, tilted sideways
1974	Kassab	Wrist support reduces muscle effort as assessed by EMG	Wrist support reduces muscle effort as assessed by EMG
1974	Herndon et al.	Reference to previous publications on CTDs	
1974	Maeda	Disorders observed in business machine operators: effects of postures and motions	
1974	Showel		Typist training methods
1974	Sidorsky		Alpha-dot keyboard developed and tested
1975	Goodwin		Cursor control keys
1975	Ayoub et al.	Changes in low frequency of the EMG as measure of muscle fatigue	
1975	Hanes		Design guidelines for keyboards and keys

Table 6.4 (*continued*)

Year	Author(s)	CTD aspects	Keying and keyboard design
1975	Einbinder		Patent to minimize finger motions by re-arranging keys, curved key rows, slanted key tops, split keyboard
1975	Engel and Granda		Cursor control devices, including keyboard
1975	Birkbeck and Beer	Light highly repetitive finger and wrist motions causal factor for CTS	High incidence of cashiers/ secretaries among CTS patients
1976	Clare		Resistance change during key travel to indicate activation (tactile feedback)
1976	Wood		Ergonomic console design
1976	Davies and Pratt	Effects of muscle contraction on manipulation	
1976	Ferguson and Duncan	29 CTD patients treated	Keyboard design and operating posture associated with CTDs
1976	Posch and Marcotte	Of 1201 cases of CTD, 36 per cent were work related	
1976	Einbinder		Patent to minimize finger motions by re-arranging keys, curved key rows, slanted key tops, split keyboard
1976	Tichauer	EMG and biomechanical procedures available to relate muscular effort and motion	

Table 6.4 (*continued*)

Year	Author(s)	CTD aspects	Keying and keyboard design
1976	Van Nes		Among keying errors, activating an adjacent (incorrect) key was the second or third most often recording error of seven error categories
1977	Hadler	CTD may be precipitated or caused by stereotyped hand use, but little formal proof is available	
1977	LeCocq		Ergonomic design of computer terminal
1977	Bequaert and Rochester		Experiments on a chord keyboard
1977	Rodbard and Weiss		Effects of reduced blood supply in arms of typists
1977	Roth et al		Ergonomic console design, specifically for reach
1978	Dorris and Purswell		Warnings needed because people are not proficient at assessing hazards
1978	Card et al.		Ergonomic evaluation of input devices including keys
1978a,b	Rochester et al. (2 publications)		Chord keyboard developed
1978	Whitaker		Moveable keys contain an arrangement to follow the fingertip positions
1978	Zapp		Keys moveable by tips of the finger, provision of hand/wrist supports
1978	Rothfleisch and Sherman	Occupational risk of CTS can be reduced by proper biomechanical and ergonomic means at the job	

Table 6.4 (*continued*)

Year	Author(s)	CTD aspects	Keying and keyboard design
1978	Tichauer and Gage	Tenosynovitis and hand overuse	
1978	Armstrong and Chaffin	Displacements of wrist and finger flexor tendons measured	Displacements of wrist and finger flexor tendons measured
1979	Cakir et al.		Details for keyboards and computer workstations in general
1979a,b	Armstrong and Chaffin (2 publications)	18 cases of CTD effects of force and wrist positions	
1979	Gopher and Eilam		New chord keyboard
1979	Kuorinka and Koskinen	17 cases of wrist syndromes related to number of work pieces handled	
1979	Luopajaervi et al.	In 152 female assembly workers, CTDs were more prevalent than in shop assistants	
1979	Waris et al.	Methods to screen for CTDs	
1980	Huenting et al.	CTDs frequent in accounting machine operators	CTDs related to posture and key operation
1980	Maeda et al.	CTDs found more in keyboard operators than in sales clerks	Recommend improvements in keyboard operator workstations, including arm support
1980	Cakir et al.		Details for keyboards and computer workstations in general

Table 6.4 (*continued*)

Year	Author(s)	CTD aspects	Keying and keyboard design
1980	Laeubli et al.	Same findings in Switzerland as in Japan Occupational CTDs related to work with office machines	Keyboard height related to body posture
1980	Zapp		Keys arranged in hand-configured array, wrist rest provided

all keys stayed on straight lines, as drawn by Sholes. Yet, human-engineered keyboards and mouse input devices were offered and marketed in increasing numbers and varieties for use with the now numerous computers, as shown in Table 6.5. Many designs divided the keyboards into one section each for the left and right hand and set the keys into hand-configured arrays, often similar to those designs offered in proposals made early in the century. Use of wrist rests, adjusting location and angulation of the keyboard, ergonomic layout of the computer workstation, and proper work habits with periods of rest from keying were becoming popular as techniques to avoid CTDs (see Table 6.5).

During the 1990s, the biomechanical strains underlying CTDs, especially carpal tunnel syndrome, were researched in detail. Their correlation to workload on the job, especially that associated with keyboarding, is well understood. As listed in Table 6.6, the specific traits of new keyboard designs to avoid overloading of their operators were tested. Proving previous concerns false, non-traditional (slit, slanted and tilted) keyboards did not require long retraining periods and did not reduce keying performance as compared to flat conventional keyboards, and use of the ergonomic keyboards resulted in better performance.

Table 6.5 Publications from 1981 to 1990 related to CTDs, keyboarding and keyboard design (Kroemer, 2001)

Year	Author(s)	CTD aspects	Keying and keyboard design
1981	Cannon et al.	CTDs related to jobs	
1981	Grandjean et al.		Split, slanted, and tilted keyboard preferred and associated with natural postures
1981	Fraser	Tenosynovitis related to excessive and repeated motions in keyboard operation; disorders due to overuse, especially excessive key force and displacement, and keying frequency	Body posture at work related to workstation layout
1981	IBM		Ergonomic guide for managers of VDT workplaces
1981	IBM		Ergonomics of VDT workstations
1981	Zipp et al.	Based on EMG findings, suggest slant and tilt of a split keyboard	Based on EMG findings, suggest slant and tilt of a split keyboard
1981	Litterick		Review of keyboard improvements
1981	Smith et al.	Musculoskeletal problems more often found with VDT operators than with control subjects	
1981	Stammerjohn et al.		Assessment of physical working conditions, particularly of keyboard height and chair

Table 6.5 (*continued*)

Year	Author(s)	CTD aspects	Keying and keyboard design
1981	Miller and Suther		Preferred adjustments for keyboard height, keyboard slope, and the VDT
1981	Tynan		A quick and easy procedure to determine ergonomic design aspects of a computer workstation
1981	Dainoff	Health issues associated with work at VDTs	Ergonomic issues associated with work at VDTs
1981	Miller and Suther		Preferred height and angle settings of display and keyboard
1981	Happ and Beaver	Strong association between fatigue and visual stress suggests the same construct as for musculoskeletal complaints	
1981	Benz et al.		Comprehensive ergonomic guide for VDT workplaces, including keyboard and key design
1981	Cannon et al.	CTS associated with use of vibrating tools	
1981	Eikelberger		Current technology allows easy custom design of keyboards, including the use of keys with "snap action" for tactile feedback
1981	Gentner		Study of finger movements while typing

Table 6.5 (*continued*)

Year	Author(s)	CTD aspects	Keying and keyboard design
1981	Hirsch		Examples of IBM human factors research on keyboards, including feedback
1981	Malt		Keys arranged for frequency of occurrence Keys are on concave surface, keyboard is split
1981	Simonelli		The arrangement of keys on a membrane keypad influenced keying time
1982	Grandjean		Postural problems related to VDU workplaces
1982	Huenting et al.		Constrained postures may be associated with physical impairments; the incidence of complaints is lowered if hands and forearms can be rested
1982	McPhee	Incidence of repetition injuries increasing in Australia; RSI related to frequency, force, posture, and time	RSI related to frequency, force, posture, and time of keying
1982	Cohen		Typing performance with membrane keyboard nearly as good with conventional keyboard
1982	Helander		Human factors guidelines for visual display terminals
1982	Norman and Fisher		Lack of justification for current keyboard size and layout; radical redesign of the keyboard promising

Table 6.5 (*continued*)

Year	Author(s)	CTD aspects	Keying and keyboard design
1982	Nakaseko et al.		Split keyboard, laterally tilted, and sections slanted improve posture and reduce fatigue
1982	Emmons and Hirsch		Comparison of keyboards of five to 18 degrees slope
1982	Dainoff	Visual and musculoskeletal complaints among VDT operators high	Linkages among stress symptoms and ergonomics of VDT workstations
1982	Suther and McTyre		Keyboard slopes between 10 and 25 degrees recommended
1982	Armstrong et al.	CTDs can be caused, precipitated, or aggravated by repeated exertions with the hand	
1982	Wagner		Review of European requirements and recommendations for NCR keyboards and their use
1982	Haider et al.		Ergonomic design of keyboarding workplaces, keyboards, and keys
1982	Helander		Office workstation design
1982	Lundborg et al.	Effects of ischemia and wrist compression on nerve function	
1982	Butterbaugh		Four key layouts had equally accurate inputs

Table 6.5 (*continued*)

Year	Author(s)	CTD aspects	Keying and keyboard design
1982	Price et al.	Operator performance measures include muscular fatigue	Operator performance measures include muscular fatigue; body posture changes with duration of typing
1983	Armstrong	CTD is an occupational illness	Design tools and tasks so that wrist displacements are avoided
1983	Jensen et al.	In 1979, more than 3000 workers' compensation claims reported for non-impact wrist disorders in 26 US States, constituting 6 per cent of all cases qualifying for compensation	
1983a	Noyes		Review of the history of QWERTY keyboard: proposals to change the design
1983b	Noyes		Characteristics, advantages and disadvantages of chord keyboards versus sequential keyboards
1983	Najjar		Review of 20 publications on keyboards since 1967
1983	AT&T Bell Laboratories		A comprehensive guide for VDT workstation ergonomics, including keyboards
1983	Clarke and Caroll		How to write user-friendly manuals

Table 6.5 (*continued*)

Year	Author(s)	CTD aspects	Keying and keyboard design
1983	Murray et al.		Voice versus keyboard control of cursor in work processing
1983	Arndt	Long recognized that repetitive motion may lead to CTDs	Posture-related complaints at VDT workstations
1983	Miller and Suther		Preferred height settings of keyboards
1983	Grandjean et al.	Reduced complaints with comfortable postures in adjustable workstations	Ability to adjust working posture. Effects of adjustable furniture on work postures
1983a–c	Kroemer (3 publications)		Ergonomics of VDT workplaces
1983	Kaplan	Typing as an occupation may contribute to the development of carpal tunnel syndrome	
1983	Chisvin	VDT operators experience more physical discomfort than other clerical employees, particularly in neck and shoulders	Postures at VDT work
1983	Gopher and Koenig		Development of a chord keyboard
1983	Benz et al.		Comprehensive ergonomic guide for VDU workplaces, including keyboard and key design

Table 6.5 (*continued*)

Year	Author(s)	CTD aspects	Keying and keyboard design
1983	Monty et al.		Acoustic feedback of key operation on several keyboards resulted in faster entry, and was strongly preferred
1983	Francas et al.		Comparison of key sets for alphabetic entry
1983	Hochberg et al.	CTDs among musicians	
1983	Zipp et al.	CTDs frequent among keyboarders	Ergonomically designed keyboard recommended, split, tilted, slanted
1983	Snyder		Recommendations for keys and keyboard: height, slope, key placement, feedback
1984	NIOSH	VDT operators report musculoskeletal strains and discomfort	Improving ergonomic conditions reduces musculoskeletal complaints
1984	Peters	Fifteen cardinal warning principles	Fifteen cardinal warning principles
1984	Burke et al.		Effects of keyboard height on performance and preference
1984	Joyce	Behavioral training resulted in reduced back, neck, and shoulder pains. Increased occurrences of stiff/ sore wrists and loss of feeling in wrists/ fingers with increased performance	

Table 6.5 (*continued*)

Year	Author(s)	CTD aspects	Keying and keyboard design
1984	Brunner and Richardson		Tactile feedback about key activation improves performance and acceptance
1984	Salthouse		High keying rates depend on perceptual chunking and simultaneous execution of digit motions
1984	Gopher et al.		High input performance on a chord keyboard can be achieved by proper spatial key arrangements, and considerations of hand symmetry
1984	Buesen		Development of a split keyboard with tilted halves
1984	Rosch		Comparison of existing keyboards in their differences in tactile feedback
1984	Life and Pheasant		Higher keyboards related to muscular effort
1984	Grandjean et al.	Preferred settings of adjustable workstations reduce complaints in neck, shoulder, and back	If adjustable, different operators set the keyboard height to widely varying levels
1984	Shute and Starr		Discomfort may be reduced with totally adjustable workstation equipment

Table 6.5 (*continued*)

Year	Author(s)	CTD aspects	Keying and keyboard design
1984	Starr	VDT use has the same physical discomfort as, but better job satisfaction than, use of paper documents for the same job	
1984	Thomas		The effects of ZH 1/618 requirements on VDT offices
1984	Cumming		Software advances make key-character assignments changeable
1984	Helander et al.	A comprehensive review of 82 studies concerning work with VDTs	A comprehensive review of 82 studies concerning work with VDTs
1984	Arndt		Key, keyboard, and workstation design
1984	Browne et al.	RSI defined as injuries caused by overload from repeated use or by maintaining constrained postures	Keyboard operators at risk
1984	Hadler	Questions CTS and job association. Lists hand and wrist disorders felt to be related to overuse	
1984	Arndt et al.	NIOSH manual for a course on health issues for VDT supervisors	NIOSH manual for a course on health issues for VDT supervisors

Table 6.5 (*continued*)

Year	Author(s)	CTD aspects	Keying and keyboard design
1984	Frey et al.		Keyboard is split into left and right halves, slanted, and tilted down to the sides; additional sets of keys are arranged further on each side, but not tilted; all key fields are located on planes sloping up away from the operator who uses the slopes to rest the arms
1985	Louis	Fatigue with conventional keyboard	Patent of a split keyboard
1985	Gaydos	Discussion of VDT issues before a subcommittee of the House of Representatives	Discussion of VDT issues before a subcommittee of the House of Representatives
1985	Koppa		Review of keypad layouts
1985a	Bartram and Feggou		Key activation affected by finger strength and mobility
1985	Nakaseko et al.	Hand posture, depending upon keyboard arrangement, affects musculoskeletal complaints	A split keyboard, both slanted in the top view and tilted in the front view, preferred over regular keyboard
1985	Gopher et al.		Investigation of coding and arrangements of chord keyboards
1985	Lynch		ANSI Standard on VDT workstations in the final stage of acceptance as an American National Standard

Table 6.5 (*continued*)

Year	Author(s)	CTD aspects	Keying and keyboard design
1985	Thomas		Promotion of good human factors in IBM products
1985	Stewart	Fatigue and discomfort from awkward postures caused by poorly located controls	
1985	Oxenburgh et al.	Duration of keyboarding and job organization related to CTD likelihood, as are psycho-social factors	
1985b	Bartram and Feggou		Keying performance depends on keyboard design
1985	Westgaard and Araas		Ergonomic improvements at work reduce CTD occurrence
1985	McKenzie et al.	Successful industry program to control CTDs based on better hand tool design, training, and management	
1985	Knave et al.	Discomfort with VDT work	Discomfort with VDT work
1985	Burry and Stoke	Repetitive strain injuries of muscles and tendons in various body parts; among the three main causes are poorly designed work station and long periods of repetitive work	Among the three main causes of repetitive strain injuries of muscles and tendons in various body parts are poorly designed work station and long periods of repetitive work

Table 6.5 (*continued*)

Year	Author(s)	CTD aspects	Keying and keyboard design
1985	Brown et al.	The Hettinger Test is able to predict whether persons are susceptible to RSI	
1986	Kirschenbaum et al.		Chorded keyboard designed to minimize physical exertion
1986	Gilad and Pollatschek		Simulation tool for keyboard layout
1986	Gopher		Two-hand keyboard developed to provide alternative to standard keyboard
1986	Silverstein et al.	Prevention strategies to avoid CTS	
1986	Rosch		New keyboard designs to replace conventional keyboards
1986	Standard Telephon Und Radio AG, ITT		Use of the ergonomic STR keyboard
1986	Hagberg and Sundalin	EMG used to assess muscle use and discomfort in word processor operators	
1986	Armstrong	CTD risk factors include repetitive exertions and body postures	
1986	Armstrong, Radwin et al.	Repetitiveness, force, mechanical stresses, posture, vibration, and temperature are CTD factors	

Table 6.5 *(continued)*

Year	Author(s)	CTD aspects	Keying and keyboard design
1986	Fine et al.	Surveillance efforts to determine causal factors of CTDs	
1986	Sauter et al.	CTDs related to keyboard use, wrist displacements, lack of wrist support or pressure on edges	Recommendations for keyboard placement and use of wrist rests
1986	Marsh	Instrument for testing sensibility of CTS patients	
1986	Statshaelsan		Detailed ergonomic recommendations for keys and keyboards
1986a-d	Fry (4 publications)	Overuse syndrome in musicians extensively reported from 1830 on, with 75 publications between 1830 and 1911 reviewed; physical signs in hand and wrist associated with overuse injury in musicians	
1986	Armstrong et al.	Repetitive motions, forceful motions, posture are CTD risks	Keyboard use exposes operators to reported CTD factors, such as repetitive motions and postures

Table 6.5 (*continued*)

Year	Author(s)	CTD aspects	Keying and keyboard design
1986	Hodges		Patent to avoid the unnatural positions of arms and wrists required at the conventional keyboard. Keys and keysets can be individually adjusted to achieve normal, natural, or restful positions for the human hands
1986	Pollatschek and Gilad		To overcome the biomechanical disadvantages of the QWERTY keyboard, a means to customize any keyboard cheaply and ineffectively is presented
1986	Seligman et al.	NIOSH found CTS associated with wrist and hand posture during typing	Adjust keyboard height relative to seat, provide wrist rests
1986	Smith	Musculoskeletal problems foremost health concern with VDT use	Keying can produce tenosynovitis and carpal tunnel syndrome
1987	Chatterjee	CTD review, risk both occupational and at leisure	Recommendations for keying work design
1987	Hocking	More than 2,200 RSI reports in about 5,000 clerical positions in Australia in the early 1980s	
1987	Ferguson	RSI epidemic in 1970s and 1980s in Australia	

Table 6.5 (*continued*)

Year	Author(s)	CTD aspects	Keying and keyboard design
1987	Richardson et al.		Investigation of various keyboards
1987a	Silverstein et al.	CTS associated with force and repetition at work	
1987	Wiklund et al.		Keyboard comparisons
1987a	Grandjean		Design of keys and keyboards
1987	Raij et al.		Motor and perceptual implications of chord keying
1987	Armstrong et al.	Relationship between repetitiveness and forcefulness of manual work, and biomechanical factors in tendonitis	
1987	Greenstein and Arnaut		Recommendations for key displacement, key force, and keying feedback
1987	Blair and Bear-Lehman	Emergence of cumulative trauma disorders like an epidemic	
1987	Louis	Course of CTD, and its treatment, predictable	
1987b	Silverstein et al.	Plant workers with hand-wrist CTD tend to transfer out of their jobs	
1987	Bleecker	The cross-section size of the carpal tunnel may be a risk factor	

Table 6.5 (*continued*)

Year	Author(s)	CTD aspects	Keying and keyboard design
1987	Feldman et al.	Surveillance and ergonomic intervention to reduce the risk of CTS in industry	
1987	Gomer et al.	EMG a useful tool to assess muscular activities and fatigue in keyboard operation	
1987	Rossignol et al.	Increased prevalence of musculoskeletal conditions with work with VDTs, apparently related to the duration of VDT work	
1987	Seror	Tinel's sign as diagnostic for CTS questioned	
1987	Ilg		Ergonomic keyboard design
1987	Sauter et al.	Pressure at edge of wrist may cause trauma	Wrist support should not present sharp pressure point, but be gently contoured and padded
1987b	Grandjean		Recommendations for VDT keyboards and workstations
1987	Herzog and Herzog		Patent of keys and keyboard sections aligned to provide proper posture of forearms and hands, and to provide accurate and unobstructed and comfortable movements of the fingers

Table 6.5 (*continued*)

Year	Author(s)	CTD aspects	Keying and keyboard design
1987	Kiser	Low-load hand activities can result in significant stress on muscles and tendons, especially with the wrist not in the straight position	
1987	Goldstein et al.	Accumulation of strain occurs in human flexor digitorum profundus tissues	
1988	Patkin	CTDs in hand-arm region	
1988	Kroemer		Survey of, and recommendations for, ergonomic means to computer workstation design
1988	Pfeffer et al.	In 1960, CTD is most frequently diagnosed, best understood, most easily treated entrapment neuropathy	
1988	Kiesler and Finholt	History of Australian RSI	
1988	Foster and Frye	RSI a case of conflicts in medical knowledge	
1988	Gopher and Raij		For mainly cognitive reasons, entry performance on separate and vertical chord keyboards for each hand was significantly faster than at a standard QWERTY layout

Table 6.5 (*continued*)

Year	Author(s)	CTD aspects	Keying and keyboard design
1988	Carr		Forces applied to keys depend on their locations
1988	Human Factors Society ANSI/ HFS 100-1988		Standards set for keys: travel 1.5–6 mm, preferred 2.0–4.0 mm; force: 0.25–1.5 N, preferred 0.5–0.6 N; feedback tactile or auditory, or both; tactile feedback preferred; reduction in key force after about 40 per cent of total displacement
1988	Rose	To avoid muscle overuse problems, China should learn from our lessons and use accumulated ergonomic knowledge for the design of a keyboard for the Chinese language	On a keyboard for the Chinese language, the maximum number of keys should be approximately 44; all key columns should be aligned rather than staggered; most frequently used keys should be located on the home row, possibly operated by the thumb; hand pronation and lateral wrist deviations should be avoided by proper key set arrangement; keyboard should be split and arranged according to existing designs in Germany, Switzerland, and Great Britain
1988	Baidya and Stevenson	EMG measurements to assess fatigue associated with wrist extensions	

Table 6.5 (*continued*)

Year	Author(s)	CTD aspects	Keying and keyboard design
1988	Hobday		To avoid lateral wrist deviation, pronation, and to follow hand shape and finger motions, the keyboard is divided into separate keysets, with the keys tilted and arranged on arcs
1988	Green et al.	Variability of joint locations	Wide range of preferred postures of keyboard operators
1988	Bjoerksten		High occurrence of neck, shoulder, and low back problems among secretaries partly explained from EMGs
1988	Molan and Sikorski	Persons with and without CTDs showed similar work behavior	
1988	Erdelyi et al.	EMG measurements to determine suitable work postures	EMG measurements to determine suitable work postures
1988	Itani et al.	EMGs are suitable means for determining muscular activities while keyboarding	EMGs are suitable means for determining muscular activities while keyboarding
1988	Krueger et al.	Repetitive work one reason for musculoskeletal complaints	
1988	Nathan et al.	No consistent association between occupational activity and slowed conduction in median nerve found	

Table 6.5 (*continued*)

Year	Author(s)	CTD aspects	Keying and keyboard design
1988	Langley		Patent of a chord keyboard that can be operated without movement of fingers from one key to another; fingers can rest on keys without activating them
1988	Putz-Anderson	Manual describes and defines CTDs, especially of the upper extremities	Management and engineering methods used in combination "to make the job fit the person, not to make the person fit the job" by redesigning the job or the tool to reduce the job demands of force, repetition and posture
1989a	Burnette and Ayoub	Definition of CTDs, prevalence, costs, pathology and etiology, treatment and rehabilitation, prevention	
1989b	Burnette and Ayoub	Computerized model to determine the CTD risk as a function of job stress and moderating factors	
1989	Ayoub and Wittels	Description of CTD injury mechanics	Workplace design, education and training, supervisory and managerial contributions are all important
1989	Hadler	No impressive data to support the contention that any upper extremity usage, within reason, is damaging	

Table 6.5 (*continued*)

Year	Author(s)	CTD aspects	Keying and keyboard design
1989	Center for Office Technology		Reports on and summarizes 21 publications on musculoskeletal research and resulting workstation design recommendations, including keyboards
1989	Morita		Keyboard design for the Japanese language with the keys divided into groups for the left and right hand, with keys on the outside located lower than those in the center; keys arranged for ease of motion, and reduction of movements; all for fast operation without excess fatigue
1989	Knight and Retter		New radically different key entry device in which the keyboard is split, and where the fingertip operates switches in different directions, from the same location; this avoids bending the wrist, flattening the hand, keeping the hands in position, repeated similar strain, and energy to operate keys
1989a	Green and Briggs		Operators must be trained in correct use

Table 6.5 (*continued*)

Year	Author(s)	CTD aspects	Keying and keyboard design
1989b	Green and Briggs	No differences between anthropometry of male CTD sufferers and non-sufferers; in contrast, significant anthropometric differences among female groups of sufferers and non-sufferers	
1989	Lockwood	CTD problems in musicians well known	
1989	Kroemer	Types and possible/likely causes of CTDs; ergonomic interventions to avoid CTDs	Ergonomic interventions to avoid CTDs
1989	Williams et al.	Exercises did not result in improvements regarding CTS	
1989	Sind		For membrane keyboards, recommends lowest possible activation force and feedback about actuation
1989	National Occupational Health and Safety Commission (Australia)	Organizational and design means to prevent overuse syndromes associated with keyboarding	Design guidance to prevent overuse syndromes associated with keyboarding

Table 6.5 (*continued*)

Year	Author(s)	CTD aspects	Keying and keyboard design
1989	Rempel et al.	Wrist tendonitis and carpal tunnel syndrome were common complaints among VDT users whose workplaces showed common ergonomic problems	Recommendations for VDT workstations
1990	Zapp		Patent of new key design; the keys are tilted instead of tapped, wrist rest is provided
1990	Faubert and Pritchard	CTDs occur in keyboard users	Reposition keys, reshape keyboard, flexible or hinged keyboards; provide alternate keyboards
1990	Hughes		Ergonomic qualities of keyboards must be ensured via standards
1990	Stewart		European Directive regarding VDUs
1990	Lachnit and Pieper		New data on speed of finger motions
1990	Ayoub	CTDs associated with extreme postures, excessive force, concentration of stress, static loading	Bank encoding console enforces unnatural and extreme postures of operator
1990	Hadler	CTDs an iatrogenic concept	

Table 6.5 (*continued*)

Year	Author(s)	CTD aspects	Keying and keyboard design
1990	Council of the European Communities		European Directive including keyboards that can be arranged to avoid fatigue and provide support for hands and arms
1990?	Helme et al.	Changes in pain sensitivity and psychometric measures of CTD sufferers	
1990	Low	Development of CTD symptoms related to duration of keyboarding	Duration of work with keyboards related to development of CTD symptoms
1990	Thomson		In a variable-geometry keyboard, an 18 degree slant and 30–60° lateral tilt minimized EMG activities and subjective discomfort
1990	Heyer et al.	VDT operators had higher prevalence of musculoskeletal symptoms than non-VDT operators	
1990	Burt (NIOSH)	Association of CTDs and keyboarding	CTDs associated with keyboarding
1990	IBM	Cumulative strain disorders may occur at VDTs	Guide to answer questions about radiation, vision, cumulative strain disorders, stress, and ergonomics at VDTs
1990	Guggenbuehl and Krueger	If the natural keying rhythm cannot be maintained, the musculoskeletal load increases	The force characteristics of the keys must suit the motor programs to avoid high musculoskeletal loads

Table 6.6   Publications from 1991 to 2000 related to CTDs, keyboarding and keyboard design (Kroemer, 2000)

Year	Author(s)	CTD aspects	Keying and keyboard design
1991	Center for Office Technology		Review of Arndt (1983): *Design for Workers*
1991	Cushman and Rosenberg		Key and keyboard design guidelines
1991	Rempel and Gerson		Actual fingertip forces influenced by force-displacement characteristics of keys
1991a	Kroemer		Ternary chord keyboard a possible competitor for QWERTY keyboard
1991	Draganova	In Bulgaria, CTD complaints by 78 per cent of data entry operators	
1991	Caple and Betts	Various factors influenced Australian RSI epidemic	
1991	Hahn et al.	CTD reduction through training and education, medical treatment, job design and placement, and workstation design	
1991	Armstrong et al.	Intra-carpal canal pressure measured with hand tasks may be sufficient to affect the median nerve	
1991	Haegg	Low-level muscle loads can affect muscle disorders depending on duration	
1991	Moore et al.	Hand CTDs explained in terms of external and internal demands, particularly regarding force, movement, repetition, and duration	

Table 6.6 (*continued*)

Year	Author(s)	CTD aspects	Keying and keyboard design
1991	Armstrong	History, causes, patho-mechanics, and control program for CTDs	
1991	Rose		Finger force applied to keys depends on hand-wrist support
1991b	Kroemer	Case studies and anecdotal evidence for CTD among keyboard operators	
1991	Apple Computer	CTD may occur by muscle or tendon overuse related to posture, repetition, and force while using computers	Guidelines on how to place keyboard, hold wrists, use light touch, take breaks
1991	Stock	Strong evidence for causal relationship between repetitive forceful work and musculoskeletal disorders	
1991	Sauter et al.		Self-reported data on musculoskeletal discomfort were collected from more than 900 VDT users, and worker posture and workstation were measured in 40 of these users; effects of ergonomic features on musculoskeletal discomfort were clearly evident from the evaluations; arm discomfort was associated with high keyboards
1992	Akagi		Key resistances were not associated with any significant typing speed differences; the high resistant linear keyboard was the least liked but showed the fewest errors

Table 6.6 (*continued*)

Year	Author(s)	CTD aspects	Keying and keyboard design
1992	Hadler	While repetitive use of the upper extremity can alter hand structure, this does not lead to increased prevalence of any musculoskeletal disease, and not to carpal tunnel syndrome; yet, repetitive use can lead to soreness; arm discomfort is real; but psychosocial aspects overwhelm workstation design aspects	
1992	Lee et al.	An evaluation of 127 exercises recommended for prevention of musculoskeletal discomfort among VDT office workers; some exercises posed potential safety hazards, exacerbated biomechanical stresses common to VDT work, or were contra-indicated for persons with health problems	
1992	Bernard et al. (NIOSH)	Evidence is provided that increasing time spent on computer keyboards is related to the occurrence of CTDs particularly for symptoms and physical findings in the hand/wrist area; for the hand and wrist, psychosocial variables were not as strong predictors as the job task variables	
1992	Grant		Patent of a keyboard that can be moved along the face of a display; the keyboard is split, the halves slanted and tilted down to the sides; the keys at the side ends of the keyboard and the cursor control keys are designed as mnemonic icons

Table 6.6 (*continued*)

Year	Author(s)	CTD aspects	*Keying and keyboard design*
1992	Flanders and Soechting		When typing a single letter, all fingers of the hand and the wrist are in motion, in a highly repeatable pattern. The simultaneous motion of the hands is not related.
1992	Soechting and Flanders		Typing is serially executed, letter by letter; when consecutive motions are done with different fingers, the motions can overlap; movement planning encompasses strings of letters
1992	Rempel, Harrison et al.	Pathophysiology, epidemiology, clinical evaluation, medical management, and prevention of work-related cumulative trauma disorders	
1992	Rempel, Dennerlein et al.		A new technique is available to measure the fingertip force during key stroke which has three distinct phases: switch compression, finger impact, and pulp compression
1992	VDT News	Various aspects of CTDs are related to physical and psychophysical conditions	Several articles that describe existing ergonomic keyboards
1993	McAtamney and Corlett	RULA (rapid upper limb assessment) is a survey method for ergonomic investigations of workplaces where work-related upper limb disorders are reported	

Table 6.6 (*continued*)

Year	Author(s)	CTD aspects	Keying and keyboard design
1993	Schoenmarklin and Marrras	Position, angular velocity and angular acceleration of the wrists of 40 industrial workers who performed highly repetitive, hand-intensive jobs of high and low risks for CTD were measured; acceleration best predicted membership to either the high or low risk group	
1993	Ranney	It is important to first identify the tissue injured, then the nature of pathology, and finally the cause; muscles may show tears or fatigue; tendons microtears and synovial thickening; and nerves hypoxia; the author describes specific steps the diagnosis, and procedures to be undertaken for prevention and treatment	
1993	Wright	New translation of B. Ramazzini's 1713 book *De Morbis Articum*	
1993	Angelaki and Soechting		Keystroke kinematics of the hand and fingers usually, but not always, vary with the typing rate; the variations in motions were mostly in the period before the key is hit
1993	Pascarelli and Kella	53 injured computer users typically had dorsiflexion of the wrist with the fingers arched, used index and middle fingers to strike keys, hit the keys hard, each habit associated with specific health deficiencies	The flat keyboard encourages placing the wrists on desk or table, often in dorsiflexion and/or ulnar deviation; strain on forearm and hand muscles, especially when splaying the fingers to reach far keys, starts a cascade of events leading to muscle damage and tendinitis

Table 6.6 (*continued*)

Year	Author(s)	CTD aspects	Keying and keyboard design
1993	Keyserling et al.	Checklist to determine the presence of ergonomic risk factors in repetitiveness, local mechanical contact stress, forceful manual exertions, awkward upper extremity posture, and hand tool use	
1993	Quill and Biers		In an on-screen keyboard, the one-line alphabetic arrangement of keys was superior to the 3-line QWERTY layout; use of a mouse was faster than using arrow keys
1994	Smutz et al.	A new measuring system to determine the effectiveness of alternative keyboards on finger force, wrist posture, operator comfort and keying performance	A new measuring system to determine the effectiveness of alternative keyboards on finger force, wrist posture, operator comfort and keying performance
1994	Carter and Banister	Possible causes of musculo-skeletal pain related to the design of workstation, chair and keyboard; operator selection, training, conditioning, posture, and rest breaks	Possible causes of musculoskeletal pain related to the design of the keyboard
1994	Wells et al.	A new system to measure the risk factors for work-related musculoskeletal disorders by combining a video image with quantitative risk information.	
1994	Fernstroem et al.	EMG activities in six fore-arm and shoulder muscles were higher when using a mechanical keyboard than when using electro-mechanical and electronic keyboards	EMG activities in six forearm and shoulder muscles were higher when using a mechanical keyboard than when using electromechanical and electronic keyboards

Table 6.6 (*continued*)

Year	Author(s)	CTD aspects	Keying and keyboard design
1994	Hales et al.	In a cross-sectional study of 533 VDT users, the hand-wrist area was most often affected by musculo-skeletal disorders; tendon related disorders were most frequent, followed by nerve entrapment syndromes; psycho-social factors play a role in the occurrence	
1994	Gerard et al.	Compared to an IBM PS/2 keyboard, muscular activities in users were reduced on a Kinesis key-board	Compared to an IBM PS/2 keyboard, muscular activities in users were reduced on a Kinesis keyboard. Learning to achieve high performance was fast on the Kinesis model.
1994	Armstrong et al.		The measured peak forces actually exerted to keys were 2.5–3.9 times the forces required to activate the keys
1994	Martin et al.		Estimated peak typing forces averaged about 4.6 times the needed key activation force
1994	Rempel et al.		Typing with keys that had a make force of 0.28 and 0.56 N did not affect the applied finger tip force or finger flexor EMG; both increased when the make force was 0.83 N
1994	Kroemer		Discussion of the shortcomings of traditional keyboards for current computer use and of possible better ways to transfer information from the human to the computer

Table 6.6 (*continued*)

Year	Author(s)	CTD aspects	*Keying and keyboard design*
1994	McMulkin and Kroemer		Five subjects learned to operate a one-hand ternary chord keyboard with 18 characters; after about 60 h of practice, their average keying was at 170 characters per minute
1994	Rudakewych et al.		A negative keyboard slope significantly improved the posture of hand, wrist, forearm, upper arm and sitting posture in all 19 subjects
1994a	Rempel and Horie	Increasing wrist extension deviations from zero while typing produced increasing pressure in the carpal tunnel	
1994b	Rempel and Horie	Resting the wrist either on a wrist rest or on the table while typing showed significantly increased pressure in the carpal tunnel over keeping the hand floating	
1994	Owen	From a US legal perspective, CTS is preventable, both on and off the job; administrative and engineering controls can reduce hazardous exposure; people should be kept productive and healthful by fitting their activities to them	
1994	Burastero et al.	A methodology to compare the results of biomechanical and performance data and subjective assessments during use of various keyboards	A methodology to compare the results of biomechanical and performance data and subjective assessments during use of various keyboards

Table 6.6 (*continued*)

Year	Author(s)	CTD aspects	Keying and keyboard design
1994	Cakir		A split and adjustable keyboard was compared to a standard keyboard both in a laboratory study and in a 6-month field study; the ergonomic design can improve postural comfort and wellbeing and reduce fatigue
1994	Faucett and Rempel	Musculoskeletal problems of VDT operators were frequent; the severity of the problems was associated with work posture, job characteristics, and length of exposure; psychosocial factors at work affected the symptoms	
1994	Marras	VDT-related CTD concerns and their prevention by ergonomics measures	
1994	Rempel et al.		During moving and pointing, mean pinch forces applied to an Apple computer mouse were about 0.5 N, but 1.4 N while dragging and almost 4 N when lifting and dragging
1995	Grant		A patent of a split keyboard, the halves slanted and tilted down to the sides; the center-mounted space bars can be activated in several directions; the keyboard sections can be set to different slope angles
1995	Moore and Garg	Six task descriptors are rated, then multipliers are assigned to each; the product of the multipliers is the Strain Index	

Table 6.6 (*continued*)

Year	Author(s)	CTD aspects	Keying and keyboard design
1995	Hedge et al.	Using a keyboard with a negative slope improved the postures of wrist and body and reduced musculo-skeletal discomfort, and was strongly preferred, in comparison to a conventional arrangement	Using a keyboard with a negative slope was strongly preferred in comparison to a conventional arrangement with positive slope
1995	Masali		The body has the natural tendency to look downward to close targets, with accompanying bends of neck and trunk, rather than look forward or upward to targets such as a computer screen
1995	Fogleman and Brogmus	Claim statistics of CTDs show a growing problem associated with the use of the computer mouse	
1995	Cakir		Split keyboards with adjustable tilt angles can improve postural and general comfort and reduce fatigue, as shown in short- and long-term studies
1995	Yoshitake		The center-to-center distance between keys can be reduced from 19 to 16.7 mm for fast typists even with large fingers, and to 15 mm for small fingers, without reducing performance

Table 6.6 (*continued*)

Year	Author(s)	CTD aspects	Keying and keyboard design
1995	Hedge and Powers		Subjects typed on a traditional 101-key keyboard placed on a 68-cm high horizontal surface (a) with and (b) without arm support, and (c) with the keyboard placed at a negative slope of 12°; ulnar deviation did not change, but dorsal wrist extension was 13° in conditions (a) and (b) and reduced to −1° with (c)
1995	Hoffmann et al.		On simulated keyboards with different key sizes and key spacings, the subjects' movement times were shortest when the spacing approximated the size of the finger pad
1995a	Bergqvist et al.	Among 353 office workers, the occurrence of muscle problems was not different in VDT- and non-VDT workers; however, the combination of VDT work of more than 20 h per week and factors such as limited rest breaks and repetitive movements was associated with risk	
1995b	Bergqvist et al.	A study of 260 VDT office workers identified factors associated with the occurrence of musculoskeletal problems; among the organizational factors were flexible rest breaks, task flexibility and overtime; among the ergonomic factors were static work posture, hand position, repetitive movements, and keyboard and VDT vertical position	

Table 6.6 (*continued*)

Year	Author(s)	CTD aspects	*Keying and keyboard design*
1996	Honan et al.		Over 4 h of intensive typing, wrist posture did not change in users of a split and tilted keyboard, but did so in users of a conventional keyboard
1996	Habes	The author defines and describes CTDs, their development due to activities on and off the job, their incidence rates and costs, and ergonomic means to control them	
1996	Sommerich et al.		When typing at preferred speed, there is no relationship between speed and key strike force, but if the typing speed is affected by external factors, strike forces tend to increase or decrease with similar changes in typing speed
1996	Martin et al.	Relatively poor correspondence between EMG data and individual dynamic finger forces may result from the fact that actual muscle load is higher than reflected in key strike force	Average peak typing forces were about 10 per cent of MVC and 5.4 times the needed key activation force
1996	Marklin and Simoneau	Split and inclined keyboards appear to reduce wrist ulnar deviation and forearm pronation from that on the conventional keyboard	Split and inclined keyboards appear to reduce wrist ulnar deviation and forearm pronation from that on the conventional keyboard
1997	Olecsi and Beaton		A new technique for continuous force-displacement measurement that assesses dynamic force applied to a key

Table 6.6 (*continued*)

Year	Author(s)	*CTD aspects*	*Keying and keyboard design*
1997	Straker et al.	Significantly greater neck flexion and head tilt occurred during work with laptop as compared to desktop computers, but other body angles showed no differences	Keying performance showed no differences between laptop and desktop computers
1997	Swanson et al.		After two days of keying on split, tilted and sloped keyboards, 50 subjects showed no significant differences in discomfort, fatigue or performance compared to work on a conventional keyboard
1997	Rempel et al.	Increasing forces exerted with the fingertip from 0 to 12 N increased carpal tunnel pressure independently from various wrist flexion and extension angles	
1997	Bernard	A critical review of epidemiologic evidence for the work-relatedness of musculoskeletal disorders (MSDs) shows, regarding keyboarding: for neck and neck/shoulder, there is reasonable evidence for a causal relationship. For shoulder MSDs, there is evidence for a positive association with highly repetitive work. For carpal tunnel syndrome, there is evidence for a positive association with highly repetitive work. For hand/wrist tendinitis, there is evidence for an association with any single factor (repetition, force, and posture), and there is strong evidence that job tasks that require a combination of these factors increase the risk for hand/wrist	

Table 6.6 (*continued*)

Year	Author(s)	CTD aspects	Keying and keyboard design
1997	Helander et al.		Provides a comprehensive overview of the contemporaneous principles and procedures used in designing the interface between computer and operator
1997	Nordin et al.	Discusses and evaluates musculoskeletal disorders of the human body in terms of epidemiology, biomechanics, psychophysiology, psychosociology, and treatment and avoidance of the disorders by ergonomic measures	Discusses and evaluates musculoskeletal disorders of the human body in terms of epidemiology, biomechanics, psychophysiology, psychosociology, and treatment and avoidance of the disorders by ergonomic measures
1998	Matias and Salvendy	The main causes of CTS are long periods of continuous typing, static and bent wrist postures, seating posture, wrist size	
1998	Martin et al.	Intramuscular and surface EMGs are in good agreement; the flexor carpi radialis and ulnaris muscles are the prime movers in typing on flat keyboards; muscle load is double as high in extensors than in flexors, muscle load increases linearly with typing speed	The flexor carpi radialis and ulnaris muscles are the prime movers in typing on flat keyboards, muscle load increases linearly with typing speed

Table 6.6 *(continued)*

Year	Author(s)	CTD aspects	Keying and keyboard design
1998	(US) National Research Council	There is a strong biological plausibility to the relationship between the incidence of musculoskeletal disorders and the causative exposure factors in high-exposure occupational settings. Specific interventions can reduce the reported rate of musculoskeletal disorders for workers who perform high-risks tasks	Specific design interventions can reduce the reported rate of musculoskeletal disorders for workers who perform high-risks tasks
1998	Greening and Lynn	Perception of vibration may be useful for early detection of onset of RSI	
1999	Carayon et al.	There are possible links between stressful psychosocial factors at work and force, repetition and posture	
1999	Marklin et al.	Split and inclined keyboards reduced the ulnar deviation from that on the conventional keyboard	
1999	Tittiranonda et al.	Eighty computer users with upper extremity musculoskeletal disorders showed improving trends in pain severity and hand function, and more individual satisfaction, when using the split, slanted and tilted keyboards than with a conventional keyboard	
1999	(US) National Research Council	Musculoskeletal disorders (MSDs) are multifactorial. The biomechanical demands of work constitute the most important contributing factors but individual, social, and organizational factors also contribute	The biomechanical demands of work constitute the most important contributing factors to musculoskeletal disorders (MSDs)

Summary

It took nearly a full century to make the general public aware that the highly repetitive action of tapping down on keys can overexert the musculoskeletal system of hands and arms. Yet research done as early as in the 1920s had already pinpointed that problem, and inventors had obtained patents for better keyboards at that time. The energy required for each key stroke decreased dramatically when electronic computers replaced mechanical typewriters, but the basic QWERTY arrangement of the keys on the keyboard itself did not change, in spite of all the well-documented shortcomings. Instead, more keys of the same kind were added, increasing the distances that fingertips must travel.

Keying-related disorders, well known to befall many typists during the first half of the century, became rampant again in the 1980s. Apparently, keyboarders were supposed to know about this; in 1997 the US Court of Appeals for the Third District decided in a class-action suit that keyboard manufacturers did not have to provide warnings about possible health risks associated with the use of their conventional keyboards.

Many kinds of improved keyboards are on the market now (in 2001), but few have tackled the apparently basic problem, namely the excessive number of repetitive key activations. Advances in technology, and the recognition of the biomechanical limitations of the human body, should lead to better-suited means of transferring information from the human to the computer.

ERGONOMIC RECOMMENDATIONS

What to do if you are not comfortable

- If you are not sure what to do in case of a fire, inquire; better still, participate in a fire drill.
- If cables and wires are in your way, make you trip, or just look untidy, have them re-arranged.
- If you often bend strongly sideways or down at your workplace to retrieve files, you may want to collect them in a file cabinet that is placed so far away that you have to walk there.
- If you feel fatigue or strain in neck, shoulder, and back after long use of the telephone, check and correct, as needed, the posture of your neck and shoulder while you hold the phone against your head. Perhaps you should use a head-carried mike and speaker.
- Do not take it lightly if you feel numbness or tingling or, worse, pain in your hands, arms or shoulders. First talk with your supervisor: you may need to re-arrange your workplace or your work habits. Furthermore, do not hesitate to get a medical diagnosis.

This feels good:

- Make sure that you know the fire alarm signal and the fastest way out.
- Cabling and wiring must be out of your way.
- It is good for you to get up often and move about, such as when looking up records in a filing cabinet placed at a distance away from your main workstation.
- If you use the telephone often or over long periods, keep it close to your ear and mouth and improve its ergonomic appropriateness by using

  - a shoulder harness or
  - better yet, a headset.

- If you do repetitive work such as keyboarding, do it correctly and take frequent breaks.
- Consider using a laptop with its smaller keyboard instead of a desktop computer.
- Try improved keyboards instead of a flat one-piece conventional keyboard.
- Make your workplace look good to you. Re-arrange furniture and devices, bring in color and light to your liking. This is likely to make you feel more at ease.
- Do not hesitate to make well thought suggestions for improvements in work procedures, equipment, tools, and appearances.

References

Ayoub, M. A. and Wittels, N. E. (1989). Cumulative Trauma Disorders. *International Review of Ergonomics 2*, 217–272.

Banaji, F. M. M. (1920). *Keyboard for Typewriters*. Patent 1,336,122. United States Patent Office.

Burnette, J. T. and Ayoub, M. A. (1989). Cumulative Trauma Disorders. Part 1. The Problem. *Pain Management, July/August*, 196–209.

Burry, H. C. and Stoke, J. C. J. (1985). Repetitive Strain Injury. *New Zealand Medical Journal 98*, 601–602.

Conn, H. R. (1931). Tenosynovitis. *Ohio State Medical Journal September*, 713–716.

Dvorak, A. (1936) *Typewriter Keyboard*. Patent 2,040,248. United States Patent Office.

Fry, H. J. H. (1986a). Overuse Syndrome in Musicians – 100 Years Ago. A Historical Review. *Medical Journal of Australia 145*, 620–625.

Fry, H. J. H. (1986b). Physical Signs in the Hands and Wrists Seen in the Overuse Injury Syndrome of the Upper Limb. Australia and New Zealand Journal of Surgery 56, 47–49.

Fry, H. J. H. (1986c). Overuse Syndrome in Musicians: Prevention and Management. *Lancet ii*, 723–731.

Fry, H. J. H. (1986d). What's in a Name? The Musician's Anthology of Misuse, in *Medical*

Problems of Performing Artists And Incidence of Overuse Syndrome in the Symphony Orchestra. Philadelphia, PA: Hanley & Belfus, 36–38; 51–55.

Hammer, A. W. (1934). Tenosynovitis. *Medical Record October 3*, 353–355.

Heidner, F. (1915). *Type Writing Machine.* Letter's Patent 1,138, 474. United States Patent Office.

Hochberg, F. H., Leffert, R. D., Heller, M. D. and Merriman, L. (1983). Hand Difficulties Among Muscians. *Journal of the American Medical Association 249*(14), 1869–1872.

Klockenberg, E. A. (1926). *Rationalization of the Typewriter and its Operation.* Berlin: Springer (in German).

Kroemer, K. H. E. (2001). Keyboards and Keying. An annotated bibliography of the literature from 1878 to 1999. *International Journal Universal Access in the Information Society.* Berlin: Springer, in press.

Lockwood, A. H. (1989). Medical Problems of Musicians. *New England Journal of Medicine 320*(4), 221–227.

Lundervold, A. J. S. (1951). Electromyographic Investigations of Position and Manner of Working in Typewriting. *Acta Physiologische Scandanavia 24* (Suppl. 84), 1–171.

Lundervold, A. (1958). Electromyographic Investigations During Typewriting. *Ergonomics 1*, 226–233.

Nelson, W. W. (1920). *The Improvements in Connection with Keyboards for Typewriters.* British Patent No. 155,446.

Pfeffer, G. B., Gelberman, R. H., Boyes, J. H. and Rydevik, B. (1988). The History of Carpal Tunnel Syndrome. *Journal of Hand Surgery 13B*, 28–34.

Schroetter, H. (1925). Knowledge of the Energy Consumption of Typewriting. *Pflueger's Archiv Gesamte Physiologische Menschen Tiere 207*(4), 323–342 (in German).

Sholes, C. L. (1878). *Improvement in Type-Writing Machines.* Letter's Patent 207,559. United States Patent Office.

Wolcott, C. (1920). *Keyboards.* Letter's Patent No. 1,342,244. United States Patent Office.

Wright, W. C. (1993). *Diseases of Workers. Translation of B. Ramazzini's 1713 De Morbis Articum.* Thunder Bay, Ontario: OH & S Press (ISBN 0-9696816-0-7).

7 Seeing and lighting

Overview

With light, we can see; without, we cannot

When the light dims, objects lose their colors; when they are in the dark, we cannot see them at all. Objects must appear bright so that we can see them in detail. Even a well-lit visual target must be at the correct distance from our eyes so that we can distinguish particulars. If our eyesight is not perfect, and as we age, it is especially important to have good lighting to carefully locate things that we must see, and we may need to refine or correct our vision with artificial lenses.

Human factor engineers know very well how to set up lighting in the office so that visual work tasks are easy to do. They also know how to choose the kind of visual targets (such as written material or computer displays) and how to locate them to make our work as easy and efficient as possible. Details are described below.

In spite of the availability of both human factor professionals and existing literature on the subject of lighting, surprisingly many fundamental mistakes are made in selecting visual targets and in setting up office illumination. This chapter gives straightforward recommendations on how to do it right.

Many muscles help us to see

Whenever we direct our sight to a new target, muscles turn our eyeballs (and possibly our head, neck, even the upper trunk) in that direction. If the object is in the midst of rapid motion, or if we are scanning something quickly, as in reading text, the eyes must move quickly as well, and this movement is achieved by muscles. As we switch from distant to close vision, muscles make the lines of sight of our two eyes converge, and other muscles adjust the thickness of the lenses in our eyes. Still further muscles adjust the pupil sizes as the visual environment progresses through different lighting levels.

Illumination, luminance and vision

Too much – or improper – light can actually reduce our ability to see. If a light source like a bright lamp shines straight into our eyes, we experience direct glare which, if strong, might make it impossible to see anything else. Daylight and lamps put visible light energy into the office. The flow of light shines onto the objects in its path; they become illuminated. Our eyes do not see the light incident on an object (called *illumination* or *illuminance*) but rather the light that is reflected from the object toward our eyes. Consider two colored surfaces, side by side, in the same flow of incident light so that they are equally illuminated; the darker surface absorbs much of the illumination and reflects only a little, while the light surface reflects most energy and absorbs little. The amount of reflected energy makes objects appear dark or light.

This reflected visible energy is called *luminance*. What determines how well we can see a visual target and distinguish its features, such as our colleague's face, a written text, or the computer display, are the contrasting luminances of the target's details. Visibility relies on the differences in luminance of the parts of the face that are brightly lit or shaded; or, put differently, on the luminance contrast between dark and light. Consider the everyday occurrence of reading written material; black letters against their pale background on either paper or the computer screen are generally easily visible to us because of the luminance contrast between dark and light.

Glare ranges in intensity; it may be simply a source of discomfort, or it may actually make seeing well a sheer impossibility. Glare occurs if sources of light (such as a blazing sunny window or an overly bright lamp – or even their reflections) are much brighter than the environment to which the eyes are adjusted. If the sun shines into your eyes, you probably cannot read the relatively dim text on your screen; if a high-energy light is reflected by the screen into your eyes, the text on the monitor does not have enough contrast for you to read it.

Feeling comfortable

What we consider comfortable is in line with the lighting engineer's design goal; we want an office environment that

- allows us to see clearly and pleasantly what we want to see;
- prevents glare and annoying bright spots in our visual field;
- pleases us in terms of contrast and colors.

> You may skip the following part and go directly to the Ergonomic Design
> Recommendations at the end of this chapter – or you can get detailed
> background information by reading the following text.

How, exactly, do we see?

While the physiology of our eyes is relatively easy to understand, the
physics of vision and lighting are more complex. Surprisingly, even
today several competing theories try to explain our color vision. Furthermore, there is still unnecessarily complicated terminology, and many of
the measuring units are metric but some are not. For detailed information
one should consult specialized texts on vision and lighting, including
those by Boff and Lincoln (1988), Boyce (1991, 1997) and Kaiser and
Boynton (1996). For ergonomic applications, several recent texts are
available – e.g., by DiNardi (1997), Konz and Johnson (2000),
Karwowski and Marras (1999), Kroemer and Grandjean (1997); Kroemer et al. (1997, 2001), Plog (2001), Post (1997) or Salvendy (1997).

Architecture of the eye

How we humans see, and how our eyes function, has been thoroughly
researched. The characteristics of human vision are well described in the
literature: e.g., Boff et al. (1986) devotes seven chapters to describing the
details of human vision.

The eyeball is a roughly spherical organ of about 2.5 cm diameter,
surrounded by a fibrous layer, called the sclera, that helps the eyeball to
retain its round shape. When a beam of light from a distant target reaches
the eye, it first passes through the *cornea*, a translucent bulging round
dome at the front of the eyeball, kept moist and nourished by tears. Then
the light runs through a chamber filled with *aqueous humor*. Behind it is
the *iris*, tissue surrounding a round opening called the *pupil*. Muscles
open and close the pupil like the aperture diaphragm of a camera, regulating the amount of light that enters the eye.

After passing through the pupil, the light beam enters the *lens*. If the
visual object is at a distance, ligaments keep the lens thin and flat, so that
incoming light rays are not bent. For close objects the ciliary muscle
around the lens makes it thicker and rounder, so that the light beams
are bent (*refracted* is the commonly used technical term) for suitable
focus. In the young and healthy eye, the cornea, aqueous humor, and
the lens refract incoming light beams so that they come to a sharp focus
on the *retina*, the innermost layer in the rear of the eyeball.

The beam of light then passes through a large space behind the lens; it is filled with the *vitreous humor*, a gel-like fluid. The light finally reaches the *retina*, a thin tissue that lines about three-quarters of the inner surface of the eyeball at its rear. The retina carries nearly 140 million light sensors that react to light energy.

There are two kinds of light sensors on the retina, named for their shape. The majority of the sensors, about 120 million of them, is comprised of *rods* which respond even to low-intensity light and provide black-gray-white vision. The other sensors are *cones*, which respond to colored light, provided it is intense enough. Cones are concentrated in a small area of the retina, called the *fovea*. It is located in the spot where a straight line, called the *visual axis*, coming through the centers of cornea, pupil, and lens, strikes the retina. Each cone contains a pigment that is most sensitive to either blue, green, or red wavelengths.

Chemical reactions in cones and rods create electrical signals that are passed along the *optic nerve* to the brain.

Muscles of the eye

Several groups of muscles adjust and move the eye. Two muscle groups are inside the eyeball for internal adjustments. One muscle pair is in the iris where it controls the size of the opening of the pupil. The other muscles are at the lens. The lens is normally held in a stretched position by the suspensory ligaments so that the eye is focused on far objects. When *accommodation* (focusing on a closer object) is necessary, the ciliary muscles overpower the ligaments and make the lens thicker. The necessary elasticity of the lens usually declines with aging, so most people in their forties or fifties need artificial lenses for seeing close objects. These manufactured lenses take over the job that the now stiff natural lenses can no longer do.

Other muscle groups are outside the eye. Six muscles attach to the outside of the eyeball controlling its movements in pitch (up and down), yaw (left and right), and roll (turning clock- or counterclockwise), all around a common center of rotation located approximately 13.5 mm behind the cornea. These muscles determine the *mobility of the eyes*. This entails primarily two kinds of movements: When one follows an object that moves along a path that stays at the same distance, the lines of sight of a pair of eyes remain parallel. This is called a conjugate movement. The other is a verging movement, where the lines of sight of a pair of eyes do not remain parallel but meet at the visual target. As it comes closer, the eyes converge; as is moves away, they diverge.

The eye can continuously track a visual target which is moving left and right (yaw) at less than 30° per second and doing so at less than 2 Hz. Above these rates, the eye is no longer able to track continuously but lags behind and then must move in jumps (saccades) to catch up to the visual target.

Other muscles that are crucial for vision include the muscles that move the head on the atlas (the top of the spine), muscles that bend the neck, even muscles that twist the trunk. Normally, they all work smoothly with the muscles of the eyes to bring objects into view, but injury or aging might influence their functioning and, hence, adversely affect vision.

Line of sight

If the eye is fixated on a point target, the straight *line of sight* (LOS) runs from the object through the pupil to the receptive area on the retina, most likely the fovea.

Until just a few years ago, it was common practice to describe the LOS direction in front of the eye (as observed from the side; see Figure 7.2) by its angle against the horizon. But this practice of using a reference external to the body obscures the fact that we unconsciously adjust not only the direction in which our eyes look but, at the same time, adjust the postures of the head, the neck and often even the upper trunk (Delleman, 2000). What we must know is the line-of-sight angle (LOSA) against the head, not against the horizon, in order to place visual targets in the workstation for both efficient and comfortable viewing. To determine the suitable line of sight (LOS), and its angle (LOSA), the ergonomist needs a reference that moves with the head, not one that is independent of head motion.

An easy-to-establish reference on the head is the ear–eye (EE) line. It runs through two landmarks of the head that can be readily seen from the side: one is the ear opening (the tragus in the ear canal, to be exact), the other landmark is the external junction of the eyelids (see Figure 7.1). (The so-called Frankfurt line, used in older texts, is more difficult to determine. It is pitched about 10° off the EE line.) The EE line is occasionally referred to as Reid's line.

The EE line allows us to define two angles (in the lateral view) as shown in Figure 7.1; the pitch P of the head against the horizon, and the pitch LOSA of the line of sight against EE. This tells us:

- How the head is held. The angle P between the EE line compared to the horizon (or to the neck, or trunk) describes the relative posture of the head. The head is said to be held erect, or upright, when the angle P between the EE line and the horizon is about 15°.
- How the line of sight is angled. The angle LOSA between the LOS and the EE line describes whether we look down, straight forward, or up to a visual target.

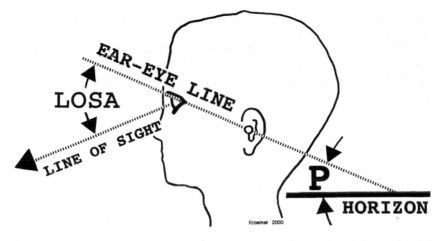

Figure 7.1 The EE line, its pitch angle P against the horizon, and the angle of the line of sight LOSA against the EE line.

The visual field

The visual field is the area, measured in degrees, within which we can see objects. To the sides, each eye can see a bit over 90°, but colors are apparent only within about 65°. Upward, the visual field extends through about 55°, with colors evident only to about 30°. Downward vision is limited to about 70°; with colors visible to about 40°.

Eyeball rotation increases the visual area to the field of fixation; it adds about 70° to the outside, but nothing in the upward, downward and inside directions because the eyebrows, cheeks, and nose stay in the way.

If the head moves in addition to the eyeballs, nearly everything in the environment can be seen, if not occluded by the body or other structures, when head and eyes are turned in that direction. Of course, mobility of the head, achieved by neck muscles, is sharply reduced in persons with a stiff neck, often experienced by the elderly. To accommodate those suffering from less mobile neck muscles, it is especially important to place visual targets close to their naturally chosen line of sight.

The naturally chosen line of sight to a target located 1 meter or less away from the eyes is, for most people,

- straight ahead, neither to the left or right, and
- between 25 and 65° below the EE line (these numbers derive from an average LOSA of 45° ± 1.65 times the standard deviation of 12°; for more details, see Kroemer et al., 2001.)

When we want to describe the best height of the visual target with reference to eye height, we must first consider the posture of the head. If the person holds the head erect (P in Figure 7.1 is about 15°), the target should be distinctly below the eyes. If one leans back in a seat (that is, as P increases), the target can be placed higher: The more we lean back, the higher it can be. (This is because the line of sight LOS, as well as the EE line, move with the posture of the head, as mentioned earlier.)

Notice that these recommendations are not fixed numbers because different people prefer various arrangements; any one individual may not be suited for a set-up that others find appropriate.

Visual fatigue

People doing close visual work, such as with a computer display, often complain of eye discomfort, visual fatigue or eye strain. While individuals' problems vary, their complaints often seem related to focusing on objects that are too far away. In other words, they are straining their eyes because their visual targets are located at a distance that differs from the minimal resting distance of accommodation, which is about 1 m away from the pupils. (This is obviously much closer than optical infinity which was previously assumed to be the automatic resting position.) As one lowers the angle of gaze, or tilts the head down, the resting point gets closer to the eyes, to about 80 cm for a LOSA of 60° downward direction. However, the resting distance increases as one elevates the direction of sight, on average to about 140 cm at a 15° upward direction. (Check Kroemer et al. (2001) for references to the related literature.) Of course, as usual, there are large variations among people in these respects. Therefore every person should be allowed and encouraged to adjust a target that is visually fixated to his/her preferred personal distance and height (Sommerich et al., 2000).

Apparently, many computer users' complaints about eye fatigue are related to the erroneous placement of monitor, source documents or other visual targets (discussed in Chapter 4) or to unsuitable lighting conditions at their workplace, as discussed later in this chapter.

It is natural to look down at close visual targets, such as a written text. Optometrists have always known this and thus placed the reading section of (bifocal or trifocal) corrective lenses into the bottom sections of the lenses.

Photometry – measurement of light

Physical measurements of light energy do not completely match its perception by humans. For example, the subjective descriptors *bright* or *dim* are not solidly related to physical measurements of illuminance or luminance.

The optical conditions of the human eye, the sensory perception of the stimuli, and our brain's processing modify the physical conditions of light. To consider this, the Commission Internationale de l'Eclairage (CIE) developed the standard luminous efficiency function for photometry in 1924 that used a model of the visual traits of the human observer. With some modifications in 1951 and 1978, the CIE 1978 standard is still valid.

The radiant flux measured in Watts (W) describes the total energy consumed for lighting and, over time, the amount shown on the electricity bill. This translates (with a few twists – see the radiometric and photometric literature for details) into the overall illumination of the office, measured in lux (lx). But luminance, measured in candela per square meter (cd m^{-2}), is the most important phenomenon for human vision; this is the light energy emitted by or reflected from a surface. Luminance is what enables us to see a wall or furniture, a written document or the computer display.

Luminance of an object is determined by its incident illuminance, and by its reflectance:

$$\text{luminance} = \text{illuminance} * \text{reflectance} * \pi^{-1} \qquad (7.1)$$

Reflectance is the ratio of reflected light to received light, in percent. The numerical value for illuminance is in lx and for luminance in cd m^{-2}. The factor π^{-1} is omitted when the following non-metric units are used: luminance in foot-lambert (fL), illuminance in foot-candle (fc).

Overall CIE, IESNA and ANSI recommendations for office illumination range from 500 to 1000 lx, even more if there are many dark (light-absorbing) surfaces in the room. But if light-emitting displays are present, such as cathode ray tubes (CRTs) on older and on many current computers, one should lower the overall illumination to between 200 and 500 lx to avoid degrading the image quality. In rooms with older flat-panel light-reflective displays, illumination of 300–750 lx is appropriate. However, with display technology changing rapidly, these recommendations may need updating soon.

How does lighting help our vision?

We can see an object only if it sends light towards our eyes. We see details well if there is strong luminance contrast between the visual target and its

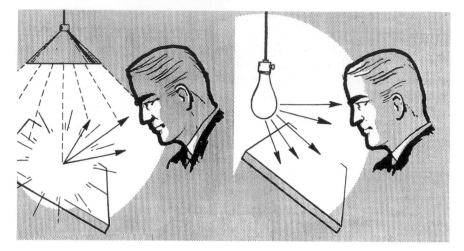

Figure 7.2   Examples of indirect (reflected) and direct glare. (Modified from Morgan et al. (1963) © McGraw-Hill.)

background. Luminance is reflected illumination that itself may come from the sun or an artificial light source (lamp, also called luminaire). If an intense light is reflected into our eyes, we experience indirect glare; if an intense light source shines straight into the eyes, we suffer direct glare (see Figure 7.2). Either type of glare can make it difficult to see what we want to see.

An accountant likes the use of *direct lighting* (when rays from the source fall directly on the work area) because it is most efficient in terms of illuminance gain per unit of consumed electrical power, but direct light can produce high glare, poor contrast, and deep shadows. The alternative is to use *indirect lighting*, where the rays from the light source are reflected in many different directions (diffused) at another surface, often the ceiling of the office, before they reach the work area. This helps to provide an even illumination without shadows or glare, but is less efficient in terms of use of electrical power. A compromise is to use a large translucent bowl that encloses the light source so that the light is scattered as it is emitted from the bowl's surface. This can cause some glare and shadows, but is usually more efficient in terms of electrical power usage than indirect lighting. Figure 7.3 shows these kinds of room lighting.

In computerized offices, the overall recommended illumination level is fairly low (200–500 lx, see above) to maintain suitable luminance of CRT displays. This illuminance may be a bit dim, especially for such tasks as reading text on a paper; it might be helpful to turn on a task light directed at the visual target which generates more luminance there without appreciably raising the overall lighting level in the office. Care must

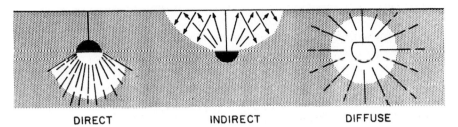

DIRECT   INDIRECT   DIFFUSE

Figure 7.3   Lamps for direct, indirect, and diffuse lighting. (From Morgan et al. (1963) © McGraw-Hill.)

be taken, however, to avoid glare by shining that light, directly or by reflection, into one's eyes.

Vision myths and truths

This section is adapted from Kroemer et al. (2001).

Myth 1: You can damage your eyesight by working in dim or glaring light, reading fine print, wearing glasses with the wrong prescription, or staring at a computer screen.

Truth: Prolonged use of the eyes under any of the stated conditions can cause eyestrain, since the eye muscles struggle to maintain a clear or unwavering focus. In addition, prolonged staring can dry the front of the eye somewhat, because it reduces blinking, which helps lubricate the cornea. But fatigue and minor dryness, no matter how uncomfortable, cannot permanently harm your vision. Of course, it makes sense to minimize the discomfort by using the following techniques:

- *Lighten up.* Age tends to cloud the lens of the eye and shrink the pupil, sharply increasing the need for luminance. So if reading strains your eyes, consider installing brighter lights or at least moving the reading lamp closer to the page.
- *Cut the glare.* Position the reading lamp so that light shines from over your shoulder, but make sure it does not reflect into your eyes from the computer monitor. Do not do read or do computer work while looking toward an unshaded window. Wear sunglasses if you are reading outside.
- *Stop and blink.* When you are working at the computer or reading, pause frequently – say, every 15 minutes – to close your eyes, or gaze away from the screen or page, and blink repeatedly. Every half hour or so, get up and take a longer break.
- *Find and maintain the right distance.* Keep your eyes at the same distance from the screen as you would from a book. If that is uncomfortable, buy a pair of glasses with a prescription designed for computer

work, or use progressive-addition bifocals, which have gradually changing power from the top to the bottom of the lens.

- *Lower the screen.* Keep the top of the screen below eye level. Gazing upward can strain muscles in the eye and neck.
- *Use a document holder.* Put a support for reading matter next to the screen, at the same distance from your eyes.
- *Clean your screen and your glasses.* Dust and grime can blur the images. If these measures do not reduce the strain from either reading or computer work, have an optometrist or ophthalmologist check if you need to start wearing glasses or have your prescription changed.

Myth 2: The more you rely on your eyeglasses or contact lenses, the faster your eyesight deteriorates.
Truth: This myth is based on the misconception that artificial lenses do the work of the eyes, which then supposedly grow lazy and weak. However, artificial lenses merely compensate for a structural defect of the eye – an improperly shaped eyeball or an excessively stiff lens – that prevents proper focusing, despite the best efforts of the lens muscles. When you wear glasses or contact lenses, the eye muscles no longer need to be tensed or relaxed (depending on whether you are farsighted or nearsighted) any more often than usual. Instead, they work just as hard as the muscles in a normal eye.

Myth 3: Eye exercises can help many people see better.
Truth: Eye exercises can help some children whose eyes have major binocularity problems, such as crossing, misalignment, or inability to converge. But claims that the exercises can help many children read better are unsubstantiated. Some maintain that eye exercises not only help one read better, but also sharpen visual acuity, boost athletic performance, and help correct numerous problems in both children and adults. Such claims are unsupported and implausible.

Color perception

Sunlight contains all visible wavelengths, but objects struck by the sun absorb some radiation. Thus, the light that objects transmit or reflect has an energy distribution different from that received. A human looking at the (transmitting or reflecting) object does not analyze the spectral composition of the light reaching the eyes; in fact, what appears to be of identical color may have different spectral contents. The brain simply classifies incoming signals from different groups of wavelengths to label them colors by experience. Human color perception, then, is a psychological experience, not a single specific property of the electromagnetic energy that we see as light.

The concept of equivalent-appearing stimuli provides a system of color measurement and specification called colorimetry (Pokorny and Smith, 1986; Post, 1997). We humans can perceive the same color even if it consists of various mixes of the three primary colors, red, green, and blue (the Young and Helmholtz theory). Therefore human color vision is called trichromatic, and the often-used CIE chromaticity diagram plots colors by the amounts of standard red, blue, and green primaries that match them.

In addition to the Young and Helmholtz theory, several other concepts try to explain human color vision, such as Hering's Opponent Colors theory and Judd's Zone or Stage theories; and other approaches are still being developed and tested. When you read the related literature you will find that the complex nature of color appearance is still not fully understood.

Esthetics and psychology of color

While the physics of color stimuli arriving at the eye can be well described, perception, interpretation, and reaction to colors are highly individual, non-standard, and variable. A person's judgment of the perceived color of a visual stimulus depends on subjective impressions experienced when viewing the stimulus, and the judgment varies with the viewing conditions and the kind of stimuli.

People believe in and describe emotional reactions to color stimuli. For example, reds, oranges and yellows are usually considered warm and stimulating. Violets, blues, and greens are often felt to be cool and to generate sensations of cleanliness and restfulness. However, likes or dislikes of certain colors and their combinations are regionally very different; travel to Asia and Europe, for example, makes this obvious. Pale colors may seem cooler than dark colors, cold colors more distant than warm colors, weak colors more distant than intense colors, soft edges of color patches more distant than hard edges, etc. Although experimental evidence on these effects is missing (Post, 1997) or controversial (Kwallek and Lewis, 1990), color schemes are often applied to work and living areas to achieve emotional responses.

The computer display

Traditional displays used a CRT to generate a picture that the human eye could perceive. Newer techniques have flatter screens, utilizing advanced light reflective and emissive displays, passive or active thin-film transistor matrices. The current ANSI/HFES 100 describes their optical properties and contains recommendations for their proper human engineering. Yet, without doubt, emerging technologies will quickly generate new ways to make electronically generated events visible to the eye. Therefore, even descriptions as in the current ANSI/HFES 100 (2001) become outdated quickly, and one must follow the technical literature closely to stay up to date.

Whatever the technology in practice, all current types of desktop computer monitors display the image on a somewhat curved screen located on the front side of a housing. The optical properties of the display determine the suitable lighting of the room in which the monitor is located, as stated earlier.

Placing the display

The eye sees what arrives at the retina: the incoming light's energies and wavelengths as distributed over the area of stimulated rods and cones. Ideally, the arriving image solely represents the electronically generated display, but it can contain the reflections of ambient light sources that are mirrored at the surface of the display. If the mirrored light is strong enough, we speak of *glare*. If it is too strong, it diminishes the apparent contrast, or color, and hence washes out the original image so that we may find it difficult, if not impossible, to discern it anymore.

Avoiding indirect glare

There are two ways to reduce the disturbing effects of *indirect glare*, such as the headlights of a tailgating car that are reflected in your car's mirrors. In the office, your view of text or other images in the computer display may be disturbed by reflected glare (perhaps without you being consciously aware of it). The less desirable means is to apply some sort of treatment to the front of the reflecting surface of your monitor. A filter is commonly used to absorb some energy of the incoming light, often in certain wavelengths (colors), but it also absorbs energy from the emitted image, which – unfortunately – reduces the luminance contrast that arrives at the eye. A related approach is to use micromesh or microlouvers that limit the directions from which stray light may fall onto the monitor surface. The disadvantage of the latter technique is that the observer must place the eyes in exactly the suitable position, forcing the head and hence upper body into a given posture. Alternately, one can use coatings on the surface or roughen the surface to minimize the reflections of incoming light – but the basic problem with all of these measures is that they just manipulate the reflections instead of eliminating the source of the disturbance.

It is much better to tackle the origin of the problem. Do not wear a white blouse or shirt that would be mirrored in the monitor. Move or turn off a lamp whose light is reflected into your eyes, or put a shield between the luminaire and the monitor (see Figure 7.4). And if it is a window whose bright surface appears on your display, then draw a

Figure 7.4 Indirect glare on the surface of a computer display can be caused by a lamp or a window.

Figure 7.5 Indirect glare caused by light that is reflected on a shiny work surface.

(dark) curtain in front of it. Of course, you often can simply reposition your workstation together with the monitor so that the bright lamp or window is to your side or somewhere else where it does not get reflected into your eyes. The same solutions work if a bright light is mirrored from your work desk as depicted in Figure 7.5. You could also make sure that there are no smooth, polished, shiny surfaces in your field of view that can act as mirrors, disturbing your eyesight.

Avoiding direct glare

The headlights of a car coming at you at night may blind you temporarily. This is an instance of direct glare, where ample light energy shines directly on your retina, overpowering the subtle image of the road that you saw as slight contrasts at low-level luminance. In the office it is probably a task light pointed at you, or the immense light coming from a window in front of you, as depicted in Figure 7.6, that optically overpowers the image presented on the monitor. What makes you regain the ability to discern the details on the display are the same measures as used to eliminate indirect glare:

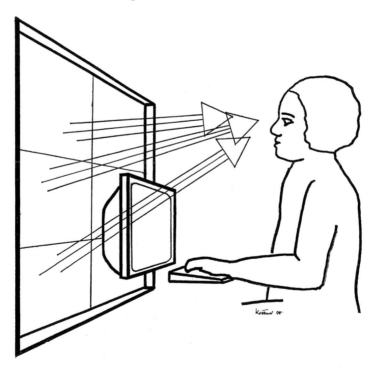

Figure 7.6   An example of direct glare. Bright light from a window shines into the eyes where its energy overpowers the weaker image coming from a computer display.

- reduce the intensity of the incoming light (turn off the lamp, draw a curtain across the window), or
- reposition yourself, your monitor and your entire workstation by about 90°.

Do not try sunglasses or eye shades.

The best ways to provide glare-free lighting from windows and luminaires can be seen in Figure 7.7. Locating light sources to the left and right sides of the operator, or overhead, is not likely to cause indirect glare, under the condition that there are no reflecting (shiny) surfaces at the workplace. Do not place a lamp or bright window in front of the

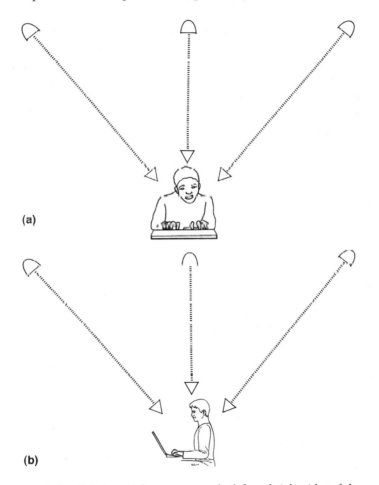

(a)

(b)

Figure 7.7   Office lighting. Light sources on the left and right sides of the operator, or located overhead, are not likely to cause glare if there are no reflecting (shiny) surfaces at the workplace. However, a light in front of the person can cause direct glare, and a light behind the person can be reflected in the display, causing indirect glare.

person because this would cause direct glare. To prevent indirect glare, avoid placing a light source behind the person because the light would be mirrored on the display surface.

In summary, the optical properties of the display and the visual properties of the human make it necessary to place the monitor with great care in relation to the human operator. The desired placement is at reading distance (about 0.5 m) from the eyes in front of the observer so that the preferred line of sight meets the center of the display. The screen should be about perpendicular to the line of sight. The line of sight should be distinctly tilted below the EE line. (Note that current laptop computers naturally come closer to this recommendation than most desktop computers.)

These stringent requirements determine the layout of office furniture and force the operator to keep eyes, and accordingly head and body, within a small space suitable for viewing the display. It is hoped that new display technology will free the human from being so strictly tethered to the monitor. Moreover, new computer input devices will ideally release the operator from the manual tie to the keyboard, as discussed in Chapter 5.

The lighting of the office, whether from windows or lamps, must be arranged so that direct glare and indirect glare are both eliminated. To this end, light sources should be carefully selected for their suitability, and carefully placed. In general, do not place them in front of or behind an operator but rather to the person's side, or overhead.

ERGONOMIC DESIGN RECOMMENDATIONS

The characteristics of human vision discussed above provide information for designing the environment for proper vision. The most important lighting concepts are:

- Proper vision requires sufficient quantity and quality of lighting, carefully arranged to provide luminance of visual targets as desired, and to avoid glare.
- What mostly counts is the luminance of an object – the light reflected or emitted from it – which meets the eye.
- The acuity of seeing an object is much influenced by strong luminance contrast between the object and its background, including shadows.
- Avoid unwanted or excessive glare. Direct glare meets the eye straight from a light source, such as the sun or a lamp shining into your eyes. Indirect glare is reflected from a surface into your eyes, such as the sun or a lamp mirrored in your computer screen.
- Use of colors on the visual target, if selected properly, can be helpful; but color vision requires sufficient light, and luminance contrast is more important than coloring.

- Colors of room surfaces affect the reflectance and hence luminance of these areas, and may affect mood and attitudes of people.
- Place the display of your computer close and low, directly behind the keyboard. Consider using a laptop instead of a desktop computer.
- The optimal lighting conditions for good vision depend on many factors including the task at hand, the objects to be seen, and the conditions of the eyes to be accommodated. Overall recommendations for office illumination range from 500 to 1000 lx, even more if there are many dark (light-absorbing) surfaces in the room. If light-emitting displays are present, such as CRTs on current computers, the overall illumination should be between 200 and 500 lx. In rooms with older flat-panel light-reflective displays, illumination of 300–750 lx is suitable.

This feels good

- Provide an overall illumination level in the office as recommended in the CIE and IESNA Lighting Handbooks, updated by the most recent ANSI/HFES 100 (see above).
- When possible, select indirect lighting, where all light is reflected at a suitable surface (at the ceiling or walls of a room, or within the luminaire) before it reaches the work area. (This helps to avoid direct and indirect glare.)
- Use several low-intensity lights instead of one intense source, placed away from the line of sight. (This avoids direct glare.)
- Place any high-intensity light sources (including windows) outside a cone-shaped range of 60° around the line of sight. (This avoids direct glare.)
- Shine a task light on your visual target (such as a paper document) if the overall illumination is too dim to generate sufficient luminance.
- Properly distribute light over the work area, which should have dull, matte, or other non-polished surfaces. (This avoids indirect glare.)
- The naturally chosen line of sight to a target less than 1 m away from the eyes is

  - straight ahead, neither to the left or right, and
  - between 25 and 65° below the EE line. (If the person holds the head erect, the target should be distinctly below the eyes, never higher.)

Notice that these recommendations are not fixed because different people prefer various arrangements – what is good for one individual may not suit everybody else.

What to do if you are not comfortable

If you feel eye fatigue or strain, check and correct, as needed

- your eyes' vision capabilities – visit an ophthalmologist or optometrist
- the distance from your eyes to visual targets
- the direction of the line of sight to visual targets
- the luminance contrast between the details of the visual target and the background (such as lines on a source document, letters on the screen)
- direct glare shining into your eyes, or reflected glare from the display or other mirroring surface.

If a window is in front of you, with the sun or bright daylight shining into your eyes:

- draw a curtain or lower blinds, or
- turn your workstation so that the window is to your side.

If your white shirt/blouse/sweater or other clothing is mirrored in your computer screen, making it hard to decipher things on the display:

- change into darker clothes.

If a task light blinds you, turn it off or turn it to the side.

If you feel fatigue or strain in neck or back, check and correct, as needed

- the posture of your neck and back, which may be affected by
    - the distance from your eyes to visual targets
    - the direction of the line of sight to visual targets.

References

ANSI/HFES 100 (2001). *US National Standard for Human Factors Engineering of Computer Workstations*. Santa Monica, CA: Human Factors and Ergonomics Society, in press.

Boff, K. R., Kaufman, L. and Thomas, J. P. (Eds.) (1986). *Handbook of Perception and Human Performance*. New York: Wiley.

Boff, K. R. and Lincoln, J. E. (Eds.) (1988). *Engineering Data Compendium: Human Perception and Performance*. Wright-Patterson AFB, OH: Armstrong Aerospace Medical Research Laboratory.

CIE Commission Internationale de l'Eclairage (1978). *CIE Publication 41, Light as a True Visual Quantity: Principles of Measurement*. Vienna: CIE.

Delleman, N. J. (2000). Evaluation of Head and Neck Postures, in *Proceedings of the XIVth Triennal Congress of the International Ergonomics Association and 44th Annual Meeting of the Human Factors and Ergonomics Society*. Santa Monica, CA: Human Factors and Ergonomics Society, 5732–5735.

DiNardi, S. R. (Ed.) (1997). *The Occupational Environment Its Evaluation and Control*. Fairfax, VA: American Industrial Hygiene Association.

IESNA Illuminating Engineering Society of North America (1993). *Lighting Handbook* (8th ed.). New York: IESNA.

Konz, S. and Johnson, S. (2000). *Work Design: Industrial Ergonomics.* (5th ed.). Scottsdale, AZ: Holcomb Hataway.

Kaiser, P. K. and Boynton, R. M. (1996). *Human Color Vision* (2nd ed.). Washington, DC: Optical Society of America.

Karwowski, W. and Marras, W. S. (Eds.) (1999). *The Occupational Ergonomics Handbook.* Boca Raton, FL: CRC Press.

Kroemer, K. H. E. and Grandjean, E. (1997). *Fitting the Task to the Human* (5th ed.). London: Francis & Taylor.

Kroemer, K. H. E., Kroemer, H. B. and Kroemer-Elbert, K. E. (2001). *Ergonomics: How to Design for Ease and Efficiency* (2nd ed.). Upper Saddle River, NJ: Prentice Hall.

Kroemer, K. H. E., Kroemer, H. J. and Kroemer-Elbert, K. E. (1997). *Engineering Physiology: Bases of Human Factors/Ergonomics* (3rd ed.). New York: Van Nostrand Reinhold–Wiley.

Kwallek, N. and Lewis, C. M. (1990). Effects of Environmental Colour on Males and Females: A Red or White or Green Office. *Applied Ergonomics 21*, 275–277.

Morgan, C. T., Cook, J. S., Chapanis, A. and Lund, M. W. (1963). *Human Engineering Guide to Equipment Design.* New York: McGraw-Hill.

Pokorny, J. and Smith, V. C. (1986). Colorimetry and Color Discrimination, in K. R. Boff, L. Kaufman and J. P. Thomas (Eds.), *Handbook of Perception and Human Performance.* New York: Wiley, 7.1–7.51.

Post, D. L. (1997). Color and Human-Computer Interaction. in M. Helander, T. K. Landauer and P. Prabhu (Eds.) *Handbook of Human-Computer Interaction* (2nd ed.). Amsterdam: Elsevier, 573–615.

Plog, B. A. (Ed.) (2001). *Fundamentals of Industrial Hygiene* (5th ed.). Itasca, IL: National Safety Council.

Salvendy, G. (Ed.) (1997). *Handbook of Human Factors and Ergonomics* (2nd ed.). New York: Wiley.

Sommerich, C. M., Joines, S. M. B. and Psihogios, J. P. (2000). Factors to Consider in Selecting Appropriate Computer Monitor Placement, in *Proceedings of the XIVth Triennial Congress of the International Ergonomics Association and 44th Annual Meeting of the Human Factors and Ergonomics Society.* Santa Monica, CA: Human Factors and Ergonomics Society, 7650–7653.

8   Hearing and sound

Overview

We want to hear – but not too much

Most humans are social creatures: we want to know the people in the office around us, get along with them, talk and work with them. (Details of social interactions in the office are covered in Chapter 2). The most efficient and important way to interact is through talking and hearing; speaking and listening are, in fact, for most of us essential work tools. However, we also want some privacy, and we certainly do not want to be disturbed or disrupted by the sounds that our colleagues make, or by noises coming from phones and other office equipment, from machinery or from outside the office. If loud sounds abound, they can make us uncomfortable, even decrease our productivity and reduce our hearing ability in the longer term.

Psycho-acoustics

Psycho-acoustics describe how we perceive sounds in relation to their physical properties. This is explained in some detail later in this chapter.

What makes us comfortable is congruent with the acoustical engineer's design goal: an office environment that

- transmits desired sounds reliably and pleasantly to the listener
- is satisfactory to the human with respect to noise
- minimizes sound-related annoyance and stress
- minimizes disruption of speech communications
- prevents hearing loss.

You may skip the following part and go directly to the Ergonomic Design Recommendations at the end of this chapter – or you can get detailed background information by reading the following text.

How, exactly, do we hear?

Sound can reach our ears via two different paths. In the office, airborne sound travels through the ear canal and excites the eardrum and the structures behind it, as described in more detail below. Sound may also be transmitted through bony structures to the head and ear, but this requires much higher sound intensities to be similarly effective.

How well we hear – distinguish sounds and understand their meaning – depends on the intensity and frequency of the sounds at their source, how they are transmitted to us (which ideally occurs without interference by other sounds), and how healthy our individual hearing organs are.

Acoustics

Acoustics is the science and technology of sound, its production, transmission, and effects. It describes the structure of sound by its physical characteristics, frequency, amplitude and duration. Psycho-acoustics establishes relations between the physics of sound and our individual perception. It uses descriptors such as pitch, timbre, loudness, noise, and understanding of speech.

Sound is first generated and then transmitted to us; thereafter we hear it, understand it and like it (or not). This simple source–path–receiver concept tells the acoustic engineer that one can manipulate sound at its source – speak a message clearly and pleasantly, for example – or prevent generation of a noise. One can also manipulate the transmission of a sound by using high-fidelity speakers if we want the signal to remain clear, for example, or reducing the energy propagation of annoying sound by materials that dam or absorb sound along its path. An example of the latter may involve sound-absorbing materials like drapery or acoustic tiles. Finally, the engineer can often improve the listener's perception through hearing aids.

Sound is any vibration (passage of zones of compression and rarefaction) through the air or any other physical medium that stimulates an auditory sensation.

For detailed information consult special publications on acoustics and audition; for ergonomic aspects you may want to check books by Berger et al. (2000), DiNardi (1997), Karwowski and Marras (1999), Kroemer et al. (1997, 2001), Plog (2001) and Salvendy (1997).

Sound, ear anatomy and hearing

Figure 8.1 shows a sketch of the ear. Sound waves arriving from the outside are collected by the *outer ear* (auricle or pinna) and funneled into the auditory canal (meatus) to the eardrum (tympanic membrane), which vibrates according to the frequency and intensity of the arriving

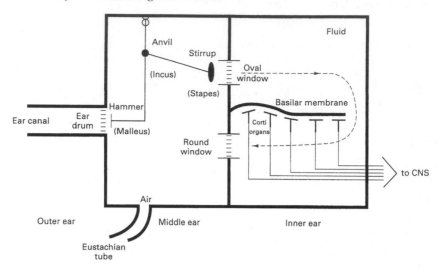

Figure 8.1   Schematic of the human ear. With permission from Kroemer et al. (2001) © Prentice Hall.

sound wave. Resonance effects of the auricle and meatus amplify the sound intensity by 10–15 decibels, dB (see below for an explanation of the dB unit) when it reaches the eardrum.

In the *middle ear*, the sound that arrived via the eardrum is mechanically transmitted by the three ear bones (the ossicles: hammer – malleus, anvil – incus, and stirrup – stapes) to the *oval window*.

Mechanical properties of the ear bones, and the fact that the area of the eardrum is much larger than the surface of the oval window, increased the intensity of the sound coming from the eardrum about 22 times when it arrives at the oval window.

Both the outer and the middle ear are filled with air. The *Eustachian tube* connecting with the pharynx allows the air pressure in the middle ear to remain equal to external pressure. However, the inner ear is filled with a watery fluid (called endolymph or perilymph) which does not need pressure adjustment.

A clogged Eustachian tube can delay the equalization of pressure between the middle ear and the environs. In a rapidly climbing or descending airplane, when you feel ear pressure, pain, or cannot hear well, you may try to open the tube by chewing gum, or by willful excessive yawning; but pumping your outer ears with your hands will not help your middle ears.

The *inner ear* contains the cochlea, a bone canal shaped like a snail shell with about two and one-half turns. Through it, sound waves move as fluid shifts from the oval window to the round window. The motion of the fluid distorts the basilar membrane that runs along the cochlea. This stimulates sensory hair cells (cilia) in the Organs of Corti which are located on the basilar membrane. The different Corti organs respond to specific frequencies. Impulses generated in the Corti organs are transmitted along the auditory (cochlear) nerve to the brain for interpretation.

The human hearing range

A tone is a single-frequency oscillation, while a sound contains a mixture of frequencies. Frequency distributions are measured in hertz (Hz), their intensities (power, amplitude, sound pressure levels) in logarithmic units known as decibels (dB). One reason for use of a logarithmic scale is that the human perceives sound pressure amplitudes in a roughly logarithmic manner; another reason involves the ease of describing the wide range of human hearing.

Infants can hear tones in the frequency range of about 16–20,000 Hz (16 Hz to 20 kHz), but few old people can hear frequencies above 12 kHz. The minimal sound pressure for hearing is 20 μPa (micropascal) in the range 1000–5000 Hz. We feel pain from about 140 Pa on, and cannot tolerate more than about 200 Pa (see Figure 8.2)

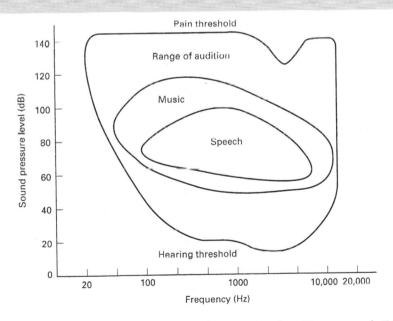

Figure 8.2   Ranges of human hearing. With permission from Kroemer et al. (2001) © Prentice Hall.

The decibel

The basic definition of the sound power level is

$$\text{Sound power level (in dB)} = 10\log_{10}(Pw_1/Pw_0) \tag{8.1}$$

with Pw being the acoustic power of the sound (usually in Watts) and Pw_0 the power at a reference level, usually set to 10^{-12} W, which is the normal hearing threshold of a young healthy ear. The decibel or dB (meaning one tenth Bell) is the unit name given to such a logarithmic ratio.

The sound pressure level SPL is also commonly measured in dB. It is the ratio between two sound pressures; one of which, the threshold of hearing (P_0), is used as reference. Since power is proportional to the square of the pressure, the equation is

$$\text{SPL} = 10\log(P^2/P_0^2) = 20\log_{10}(P/P_0) \text{ in dB} \tag{8.2}$$

where P is the root-mean-square (rms) sound pressure for the existing sound and P_0 the threshold of hearing, 20 μPa.

With these values, the dynamic range of human hearing from 20×10^{-6} to 200 Pa is

$$20\log_{10}[200/(20 \times 10^{-6})] = 140 \text{ dB} \tag{8.3}$$

Ranges of sound pressure levels are shown in Figure 8.2.

Pychophysics of hearing

While physical measurements (such as in dB) can explain acoustical events, persons interpret them and react to them in very subjective manners – you may find certain sounds loud or soft, for example, or attractive or noisy; your neighbor or colleague may quite disagree, however, with your interpretation.

The experienced *loudness* of a tone (or complex sound) depends on both its intensity and its frequency. Compared to the intensity at 1000 Hz, at lower frequencies the sound pressure level must be increased to generate the feeling of equal loudness. For example, the intensity of a 50-Hz tone must be nearly 100 dB to sound as loud as a 1000-Hz tone with about 60 dB. However, at frequencies in the range of about 2000–6000 Hz, the intensity can be lowered and the sound still appears as loud as at 1000 Hz. Yet, above about 8000 Hz, the intensity must be increased again, even above the level at 1000 Hz, to sound equally loud. The equal-loudness contours (called phon curves) are shown in Figure 8.3.

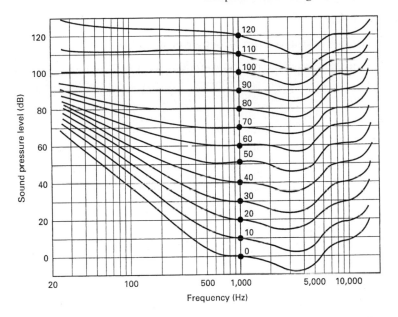

Figure 8.3 Phon curves are the lines along which combinations of sound pressure level and frequency are perceived as having the same loudness as at 1000 Hz. With permission from Kroemer et al. (2001) © Prentice Hall.

These perceptions of equal loudness indicate that there are non-linear relationships between pitch (the perception of frequency) and loudness (the perception of intensity). Timbre is even more complex because it depends on the changes in frequency and intensity over time.

Filters that are applied to sound-measuring equipment imitate the differences in human sensitivity to tones of different frequencies. These filters, today often known as software, correct the physical readings to reflect what the human perceives. Different filters have been used, identified by the first letters of the alphabet. The A filter is most often used because it is close to the human's response at 40 dB. A-corrected decibel values are identified by the notation dBA or dB(A).

Acoustic facts

This section is adapted from Kroemer et al. (2001).

Directional hearing. Humans can determine from whence a sound is coming by using the difference in arrival times or intensities at their two ears.

Distance hearing. The ability to determine the distance of a source of sound is related to the fact that sound energy diminishes with the square of the distance traveled, but the human perception of energy depends also on the frequency of the sounds, as just discussed. Thus, a source of sound appears more distant when it is low in intensity and frequency and appears closer when it is high in intensity and frequency.

Difference and summation tones. Two tones that are sufficiently separated in frequency are perceived as two distant tones. When two such tones are very loud, one may hear two supplementary tones. The more distinct tone is at the frequency difference between the two tones; the quieter tone is the summation of the original frequencies. For example, two original tones, at 400 and 600 Hz, generate a difference tone at 200 Hz, and a summation tone at 1000 Hz.

Common-difference tone. When several tones are separated by a common frequency interval, one hears an additional frequency based on the common difference. This effect explains how one may be able to hear a deep bass tone from a sound system that is physically incapable of emitting such a low tone.

Aural harmonics. One may perceive a pure sound as a complex one because the ear can generate harmonics within itself. These subjective overtones are more pronounced with low- than with high-frequency tones, especially if these are about 50 dB above the threshold for human hearing.

Intertone beat. If two tones differ only slightly in their frequencies, the ear hears only one frequency, called the intertone, that is halfway between the frequencies of the original tones. The two tones are in phase at one moment and out of phase at the next, causing the intensity to wax and wane; thus, one hears a beat.

Doppler effect. As the distance between the source of sound and the ear decreases, one hears an increasingly higher frequency; as the distance increases, the sound appears lower. The larger the relative velocity, the more pronounced is the shift in frequency. The Doppler effect can be used to measure the velocity at which source and receiver move against each other.

Concurrent tones. When two tones of the same frequency are played in phase, they are heard as a single tone, its loudness being the sum of the two tones. Two identical tones exactly opposite in phase cancel each other completely and cannot be heard. This physical phenomenon (called destructive interference or phase cancellation) can be used to suppress the propagation of acoustical or mechanical vibrations.

Noise and hearing impairments

Sounds of very high intensity, especially if they last long, regularly result in temporary or permanent hearing loss. There are two types of injuries. Short-duration sound of high intensity, such as produced by an explosion, can damage any or all of the structures of the ear, in particular the hair cells in the Corti organs, which may be torn apart. This results in immediate, severe, and permanent hearing loss, called a permanent threshold shift, PTS.

Sound levels of less than about 100 dB often cause a short-term hearing loss, measured as temporary threshold shift (TTS). During quiet periods, hearing returns to its normal level. Repeated exposures to sounds that cause TTS may gradually bring about a permanent threshold shift (PTS), just as a loud explosion can do, called a noise-induced hearing loss (NIHL) – see below for a discussion of noise. Often, the damage is confined to a special area on the cilia bed on the cochlea, related to the frequency of the sound. With continued sound exposure, more hair cells are damaged, which the body cannot replace; also, nerve fibers from that region to the brain may degenerate, accompanied by corresponding impairment within the central nervous system.

Sound level, frequency, and duration of exposure (singly or repeatedly) are the critical descriptors for sounds that can damage hearing. Sound levels below 75 dBA apparently do not produce permanent hearing loss, even at about 4000 Hz, where people are particularly sensitive. At higher intensities, however, the amount of hearing loss is directly related to sound level and its duration. In the USA, current OSHA regulations allow 16 hours of exposure to 85 dBA, 8 hours to 90 dBA, 4 hours to 95 dBA, etc. Other countries have different regulations and limits in terms of allowable exposure to noise. To illustrate, in Europe, 8 hours at 90 dBA are allowed, but 4 hours at 93 dBA, or 16 hours at 87 dBA.

Simple subjective experiences can indicate the existence of hazardous sound exposure. Indications of dangerous sound environment are:

- hearing a sound that is appreciably louder than conversational level;
- hearing a sound that makes it difficult to communicate;
- experiencing the sensation of ringing in the ear (tinnitus) after having been in the sound environment;
- experiencing the sensation that sounds seem muffled after leaving the noisy area.

Permanent hearing loss may be from mild to profound. Since we know its nature, it can be prevented or, after it occurs, alleviated by hearing aids. However, unfortunately, if the underlying nervous structure is damaged, the hearing loss is not treatable under the auspices of current technology and expertise. NIHL increases most rapidly in the first years of exposure; after many years, it levels off in the high frequencies, but continues to worsen in the low frequencies (National Institutes of Health, 1990).

> In Western countries, with their specific noises, noise-induced hearing loss usually starts in the range of 3000–6000 Hz, particularly around 4000 Hz. Then it extends into higher frequencies, culminating at about 8000 Hz. Yet, reduced hearing at (and above) 8000 Hz is also brought on by aging. This may make it difficult to distinguish between environment- and age-related causes of NIHL.

Noise in the office

Dripping water makes single, short tones of low intensity: we consider this noise just as we consider the loud, lasting, complex sounds of the neighbor's music to be noise. Noise can be measured and described in physical units, but in the office its effects usually are psychological and subjective and depend on many circumstances.

> Noise is defined as sound that is unwanted, objectionable, annoying or unacceptable to a person; it is often but not always loud.

Noise in the office can

- make it difficult to hear and understand wanted sounds, especially spoken communications;
- create negative emotions, including feelings of surprise, frustration, and anger;
- interfere with a person's sensory and perceptual capabilities and hence degrade task performance;
- temporarily or permanently reduce our hearing capability if the noise has more than about 80 dB, even if one considers it pleasant as when listening to loud music.

Audiometry and understanding speech

Assessment of hearing ability, especially its loss, includes measures of the auditory thresholds (sensitivity) at various frequencies. Pure-tone audiometry, done by ophthalmologists and optometrists, is often combined with measures of understanding speech, which is covered below.

Fred tosses the folder onto his desk and pushes his chair back with an exasperated sigh. He glances at his watch. It is already 4:45 p.m., his carpool leaves in 10 minutes, and he really should finish the budget reports before he leaves today. Trouble is, however, he just cannot seem to concentrate lately. As a financial analyst in a mid-sized bank, he always has a full workload, and this week has been especially busy so far. But he has only finished about half of his normal work. Fred sighs again as he picks up the folder he tossed on his desk; with some annoyance, he shoves the folder into his briefcase and thinks of how much of this work he will need to do at home just to catch up. Just then, a volley of hammering interrupts his thoughts. The bank is expanding, and the office is being reconfigured to accommodate new staffers. All week long, hammering, pounding, and construction crew voices have echoed through the hallways. Fred despises all this noise and knows it is adversely affecting his ability to focus on his work. He picks up his heavy briefcase and trudges downstairs to meet his carpool.

Forty minutes later, he walks into the front door of his house in the suburbs. His teenage son is at the kitchen table, doing his homework, with a radio blasting music next to him and the television blaring in the background. Yet he is so absorbed in his work that he barely registers his father's arrival. Fred shakes his head, bemused. He himself can barely concentrate when even the slightest noise disrupts him. His son, however, seems to have no problem focusing even with pulsating music pounding right next to him and the television blaring in the background. How does he do it? And why can Fred not?

Voice communications

Much of the exchange of information in the office is by speech; either with other people in the same room, or by telephone with people who we cannot see.

Voice communications use frequencies from about 200 to 8000 Hz, with the range of about 1–3 kHz most important for understanding. Men use more of the low-frequency energy than women.

Difficulty in understanding language is often associated with problems of differentiating speech sounds in their high frequency ranges. In many languages, consonants consist of high frequencies that, unfortunately, have less speech energy than vowels and therefore are more difficult to understand, especially for people with noise-induced hearing loss and older persons. Also, other sounds such as background noise, competing voices, or reverberation may interfere with the listener's ability to receive information and to communicate. Consequently, important informational content of speech may be unclear, unusable, or inaudible.

The ability to understand the meaning of words, phrases, sentences and whole communications is called *intelligibility*. This is a psychological process that depends on acoustical conditions. For satisfactory communication of most voice messages in noise, 75 per cent intelligibility at minimum is required. Perhaps you have at some point experienced that a discussion on the telephone was difficult to understand due to background and office noise around you; yet a conversation with your colleague just moments later was easy to comprehend, even though both dialogues appeared to be conducted at the same level of sound. Direct, face-to-face communication provides visual cues that enhance speech intelligibility, even in the presence of background noise. This is because non-verbal visual cues combine with speech to help direct us to the meaning of the messages given. Indirect voice communications, such as by voice-only telephones, lack the visual cues and thus might be less intelligible.

Speech-to-noise ratio

The intensity level of a speech signal relative to the level of ambient noise fundamentally and profoundly influences the intelligibility of speech. The commonly used speech-to-noise ratio (S/N) is really not a fraction but a difference: for a speech of 80 dBA in noise of 70 dBA, the S/N is simply + 10 dB. With an S/N of + 10 dB or higher, people with normal hearing should understand at least 80 per cent of spoken words in a typical broadband noise. As the S/N falls, intelligibility drops to about 70 per cent at 5 dB, to 50 per cent at 0 dB, and to 25 per cent at −5 dB. People with noise-induced hearing loss may experience even larger reductions in intelligibility, while persons accustomed to talking in noise do better.

Team-member efficiency is often impaired when noise interferes with voice communication. If this occurs, the time required to convey information is increased through the necessity of more deliberate verbal exchanges. Not only is this annoying, but it slows us down when fast action is needed, and, even more importantly, can result in increased human error due to misunderstandings.

Clipping

Filtering is often used in the transmission of speech by telephone, often called clipping because it cuts out certain frequencies or amplitudes. Here, the following findings apply:

- *Frequency clipping* affects vowels if the clipping occurs below 1000 Hz, but has little effect if above 2000 Hz. *Peak clipping* is usually not critical if below 600 Hz or above 4000 Hz, but *center clipping* is highly detrimental, particularly in the ranges of 1000–3000 Hz.

- In *amplitude clipping*, cutting the *peaks* primarily affects vowels and reduces the quality of transmission in general, but is not usually a great problem. Surprisingly, peak clipping and then re-amplifying improves the perception of consonants, which carry most of the message. *Center clipping*, in contrast, garbles the message because it primarily affects consonants.

Improving speech communication

Speech communication is explicitly dependent on the frequencies and sound energies of any and all interfering noise. Several techniques exist to predict speech intelligibility based on narrowband noise measurements: examples include the Articulation Index, the Speech Interference Level or the simpler Preferred Speech Interference Level (please see the literature for more specific information). Using these measures, one can decide what kind of interfering sound must be eliminated, and whether the speech must be improved.

Speech consists of five major components: the *message* itself, the *speaker*, the *transmission*, the *environment*, and the *listener*.

1. The message becomes clearest if its context is expected, its wording is clear and to the point, and the ensuing actions are familiar to the listener.
2. The speaker should use common and simple vocabulary with only a limited number of terms. Redundancy can be helpful (e.g., Boeing 747 jet). The speed should be slow rather than fast. Phonetically differentiated words should be used. The International Spelling Alphabet is shown in Table 8.1.

Table 8.1 International spelling alphabet

a: alpha	n: november
b: bravo	o: oscar
c: charlie	p: papa
d: delta	q: quebec
e: echo	r: romeo
f: foxtrot	s: sierra
g: golf	t: tango
h: hotel	u: uniform
i: india	v: victor
j: juliet	x: xray
k: kilo	y: yankee
l: lima	z: zulu
m: mike	

3. The transmission, for example by telephone, should be of high fidelity which has little distortion (such as by clipping) in frequency, amplitude, or time.
4. The environment should be acoustically designed; specifically, it should have no noise or reverberation that would interfere with the messages. Two main strategies can be employed:

   - *Avoid noise generation.* This is the first, fundamental, and most successful strategy. Avoid or reduce the generation of unwanted sound by proper office layout, good design of equipment, quiet flow of air, and by suitable behavior such as keeping voices low. Active noise attenuation is a promising new technique in which sounds become physically erased by instantaneously generated counter-sounds that are of the same frequency and amplitude but of the opposite direction (180° off-phase). Currently, this works best at frequencies below 1 kHz.
   - *Impede noise transmission.*The second strategy is to impede the transmission of unwanted sound from the source to the listener. In occupational environments, one can encapsulate the noise source, put sound-absorbing or damming surfaces in the path of the sound, or physically increase the distance between source and ear. Another solution is to wear a hearing protection device (HPD) (plug, muff, helmet) that reduces the annoying subjective effects of sounds and avoids auditory harm. Of course, in general, if a condition exists in an office that necessitates wearing a HPD, we should first and foremost examine the environment and, if at all feasible, remedy the condition.

5. The final component of speech is the listener. The listener's ability to understand the message is, of course, affected by existing noise – which should be eliminated. In unusual circumstances it may be necessary to wear a special HPD that is penetrable by desired frequencies and to employ modern electronic devices that amplify certain bandwidths of sound depending on the existing amplitude and (passively or actively) suppress ambient noise.

Also, the listener's hearing ability affects his or her facility in understanding a message. A number of devices are on the market that improve a person's hearing.

Improving hearing ability

Many people, especially as they get older, experience reduced hearing ability, predominantly in the higher frequency ranges. Devices to aid impaired hearing can offer a significant improvement in understanding speech, especially when these devices utilize electronics to amplify certain bandwidths of sound – depending on the existing amplitude – and filter out ambient noise (similar to HPDs mentioned above).

In reality, however, many people with impaired hearing become very frustrated when trying to use these hearing aids. Perhaps the device was not properly selected in the first place, or it was not replaced by a new and better one after periods of use (technology, after all, relentlessly marches on, and hearing aids continue to improve). Some hearing aids are difficult to use – and we must face the fact that our motor skills often get worse over the years. Barnes and Wells (1994) listed the main triggers of dissatisfaction of American hearing-aid users:

- must remove aid to use the telephone;
- could not maintain proper volume;
- was not alerted of battery running low;
- troublesome battery changing;
- experienced discomfort while wearing aid;
- device whistled at certain volumes.

Clearly, better ergonomic and acoustic design could overcome the great resistance that many people feel when it comes to wearing hearing aids. We also should make them less visible – let us just admit that vanity and pride influence us at any age – and change the general attitude so that we can accept devices that improve audition as easily as we embrace those that refine vision.

While serving as US President, Bill Clinton publicly announced that he was wearing a hearing aid, and this publicity helped to remove the common aversion to (admitting to) using a hearing aid.

Music while we work?

Music is probably one of the oldest art forms and earliest expression of human emotion. Music has long accompanied activities; consider workers singing during field labor or soldiers marching to rhythmic music. Effects of music on industrial work were observed early in the twentieth century. Yet, even today, little is known systematically about the psychophysical consequences of different kinds of music and rhythm and their effects on productivity over extended periods of time. Personal experience indicates that music can help exercisers last longer in an aerobics class or run an extra lap or two, and there are even studies on whether or not we eat meals more quickly in pace to an invigorating beat (apparently, we do), but we still do not have much conclusive research on music's effect on office workers.

Background music is like acoustical wallpaper in shops, hotels, waiting rooms, and the like. It is meant to create a welcoming atmosphere, to relax customers, to reduce boredom, and to cover other disturbing sounds. Generally, in such settings the music's character is subdued, its tempo intermediate, and vocals are avoided. It may produce a monotonous environment for those continuously exposed to it, while it can appear pleasant to the transient customer.

Music while you work is in many respects the opposite of background music. It is not continuous but programmed to appear at certain times, it has varying rhythms with vocals, and it may contain popular fashionable tunes. It is meant to break up monotony, to generate mild excitement, and to provide an emotional push during demanding physical effort or to relieve drudgery in a dreary impersonal environment. While improved morale and productivity have been reported in many circumstances, a clear linking of the underlying arousal theory and the specific components of the music is difficult (Konz and Johnson, 2000; Kroemer and Grandjean, 1997). Thus, the musical content, the rhythm, the loudness, the timing, and the selection for certain activities, environments, and listener populations are still a matter of art rather than science.

Few if any scientific statements exist about the type (or even value) of music that a person may select for individual listening, so we will avoid making any general statements at all. Music at work is a highly individual preference. The beauty of such music, often delivered via personal radios or CD players and headphones, is clearly in the ears of the beholder.

ERGONOMIC DESIGN RECOMMENDATIONS

This feels good

- The overall sound level in the office should be between 50 and 75 dB, best near 65 dB.
- For persons doing work that requires intense concentration over long periods of time, the existing sound level should not change appreciably or dramatically; most people find changes invigorating, but not if these changes are abrupt, startling or otherwise unpleasant. (However, what may be abrupt, startling or unpleasant depends on the circumstances and is also highly individual.)
- Some reverberation in the room is actually desirable – specifically, the reflection of sound at hard surfaces (floor, walls, ceiling, windows, furniture) – because it makes speech sound alive and natural.

What to do if you are not comfortable.

If it is too loud for you:

- Eliminate the sound at its source:
  - replace noisy machines with quieter equipment
  - turn down the sound level of the ringer on telephones
  - ask your co-workers for quieter behavior

- Reduce your exposure to the sound:
  - move to a different office
  - place a noisy piece of equipment outside your room
  - encapsulate the source of sound
  - if there is too much reverberation in the office, soften hard surfaces that otherwise reflect the sound; use for example drapes, carpets, acoustic tiles and the like to dampen or absorb sound

If all else fails:

- mask offending sounds by creating a "sound curtain" such as by playing music that is pleasant to you (perhaps through your personal ear phones) or have a generator of pink noise (or white noise) installed
- use personal hearing protectors (muffs or plugs)

If it is too quiet for you:

- play music
- have other people move into your office
- if your office is "sound dead" because there is too little reverberation, consider removing some drapery, carpets, acoustic tiles and similar sound-deadening materials; bare floors and walls, large pictures under glass, or windows with hard surfaces that reflect sounds.

References

Barnes, M.E. and Wells, W. (1994). If Hearing Aids Work, Why Don't People Use Them? *Ergonomic Design April*, 18–24.

Berger, E. H., Royster, L. H., Royster, J. D., Driscoll, D. P. and Layne, M. (2000). *The Noise Manual* (5th ed.). Fairfax, VA: American Industrial Hygiene Association.

DiNardi, S. R. (Ed.) (1997). *The Occupational Environment: Its Evaluation and Control*. Fairfax, VA: American Industrial Hygiene Association.

Konz, S. and Johnson, S. (2000). *Work Design: Industrial Ergonomics*. (5th ed.) Scottsdale, AZ: Holcomb Hataway.

Karwowski, W. and Marras, W. S. (Eds.) (1999). *The Occupational Ergonomics Handbook*. Boca Raton, FL: CRC Press.

Kroemer, K. H. E. and Grandjean, E. (1997). *Fitting the Task to the Human* (5th ed.). London: Francis & Taylor.

Kroemer, K. H. E., Kroemer, H. B. and Kroemer-Elbert, K. E. (2001). *Ergonomics: How to Design for Ease and Efficiency* (2nd ed.). Upper Saddle River, NJ: Prentice Hall.

Kroemer, K. H. E., Kroemer, H. J. and Kroemer-Elbert, K. E. (1997). *Engineering Physiology: Bases of Human Factors/Ergonomics* (3rd ed.). New York: Van Nostrand Reinhold–Wiley.

National Institutes of Health (NIH) Consensus Development Panel (Ed.) (1990). Noise and Hearing Loss. *Journal of the American Medical Association 263* (23), 3185–3190.

Plog, B. A. (Ed.) (2001). *Fundamentals of Industrial Hygiene* (5th ed.). Itasca, IL: National Safety Council.

Salvendy, G. (Ed.) (1997). *Handbook of Human Factors and Ergonomics* (2nd ed.). New York: Wiley.

9 Office climate

Overview

We are not cold-blooded

The body temperature of quite a few animals, such as fish and salamanders, simply changes with their environment: when it is cold, they are cold, and they are warm when it is warm outside. (The term *cold-blooded* to describe these creatures is thus somewhat misleading.) We humans maintain a rather constant temperature of about 37°C in the core, especially in the brain and chest cavity of our bodies, regardless of the outside climate.

Heat exchanges

Keeping a constant core temperature is a complex task for the human body's temperature control system. The body itself generates heat energy, and at the same time exchanges energy with the environment. When our surroundings are cold, we want to prevent excessive heat loss, which is mostly done by choosing suitable clothing. When our environment is hot, our body must achieve an outward flow of heat to prevent overheating. This can be rather difficult to do, because heat always flows from the warmer to the colder. There are technical ways to help create that outward heat flow, with the most complete – albeit also most expensive – solution involving climatizing our environment: controlling its temperature, humidity and air movement.

Thermodynamics

The exchanges of heat energy between the environment and the body follow the thermodynamic processes of radiation, convection, conduction and evaporation. Their effectiveness depends on several conditions, especially the insulating properties of the clothing we wear and the energy level of the work we perform.

Feeling comfortable

For our wellbeing and comfort, the temperature difference between exposed skin and the environment is very important, but humidity and airflow also play major roles – as do our attitude and level of acclimatization. Within reasonable limits, given suitable clothing, the human can function in both warm and cool environments.

You may skip the following part and go directly to the Ergonomic Design Recommendations at the end of this section – or you can get detailed background information by reading the following text.

What, exactly, is a climate?

Some people use the term office climate when referring to how the persons in an office treat each other, work with each other, and especially how supervisors and the supervised get along with each other. These are very important (psychological and sociological) features, covered in detail in Chapter 2, but we bring them up here only to differentiate them explicitly from physical climate. Physical climate can be described in two ways: one is to scientifically measure the present air temperature, humidity, and air movement which, together with energy generated and exchanged, objectively describe the environment's thermodynamics. The other aspect, and the more important one for the people in the office, is how each individual feels about these thermodynamic conditions; whether or not they are subjectively perceived as comfortable.

How the climate in the office affects us

Our body's temperature control system tries to keep us at an ideal temperature, so that we feel neither overly chilled nor too warm. Fortunately, in most offices the climate conditions do not change even as the outdoor environment shifts and changes, so the task of our body's thermo-regulatory system is fairly simple. To understand how the thermo-regulatory system works helps engineers to design appropriate climate control systems and allows us to select suitable settings, clothing and behaviors to be physically comfortable with the climate in our office surroundings. Accordingly, we will discuss the basics of the human body's thermo-regulatory system below.

Our body generates heat from the energy that we absorb from food and drink. The heat energy is circulated throughout the body by the blood. We exchange heat with our environment by gaining heat in a hot climate and losing it in the cold. That exchange takes place mostly through our skin. How much energy we exchange with our surroundings depends on the surface of skin participating in the exchange, especially of skin that is exposed. Clothing helps insulate the skin surface from heat transfer.

Thus, whether or not we feel comfortable depends on many features, some of which are easy to measure (such as temperature), while others are subjective and depend on how our body functions (such as metabolism and circulation), on how we dress, on how we move about, and on what season it is.

If our body's core temperature were to change about 2° from its 37°C norm, our body's functions and resultant task performance would suffer severely. However, in a sheltered office environment, the human thermoregulatory system has no problem keeping the core close to 37°C. Although the core remains relatively constant, at the skin, there are major temperature fluctuations from region to region. For example, the toes may be at 25°C, legs and upper arms at 31°C, and the forehead at 34°C all at the same point in time; for most of us, this feels comfortable.

The German physicist Daniel Gabriel Fahrenheit (1686–1736) worked with thermometers and set the temperature of a mix of ice and water to 32°, and the temperature of boiling water 180° higher, at 212°. The Fahrenheit scale was useful for scientists, but in 1742 the Swedish astronomer Anders Celsius (1701–1744) suggested a metric scale with the freezing temperature of water set to zero and its boiling temperature to 100°. This centri-grade scale was called Celsius scale by international agreement in 1948 and is used in all countries except in the USA. For conversion of temperatures' degree values from one scale into the other one must consider the different settings for freezing and boiling temperatures as well as the number of degrees between freezing and boiling of water:

Fahrenheit to Celsius :   [degF − 32] [5/9] = degC

Celsius to Fahrenheit :   [9/5 degC] + 32 = degF

The energy balance

We can use a simple equation to describe the energy exchange between inputs to the body and outputs from it as

$$I = M = H + W + S \qquad (9.1)$$

where I is the energy input via food and drink which the body transforms into metabolic energy M. The energy quantity M divides into the heat H that must be dispelled to the outside, the external work W done, and the energy storage S in the body.

You may want to check books on human metabolism for more detailed information, or ergonomic texts such as by DiNardi (1997), Karwowski and Marras (1999), Kroemer et al. (1997, 2001), Plog (2001) and Salvendy (1997).

Assuming for convenience that the quantities I, W and S remain unchanged, one can concentrate on the heat energy exchange with the thermal environment according to

$$H = I - W - S \qquad (9.2)$$

The system is in balance with the environment if all metabolic energy H is dissipated to the environment. However, the case is complicated by the fact that the body often receives additional heat energy from warm surroundings – or may lose too much heat in a cold room.

Transmitting energy by radiation, convection, conduction and evaporation

The human thermo-regulatory system interacts with the environment. In a comfortable climate, as in most offices, the task is to dissipate heat energy generated by the body's metabolism. This is thermodynamically easy in a cold environment, but we may feel cool; dissipating heat energy is more difficult in warm environs where we may feel hot.

Energy exchanges with the environment

Energy is exchanged with the environment through radiation (R); convection (C), conduction (K) and evaporation (E).

Heat exchange by radiation, R, is a flow of electromagnetic energy between two opposing surfaces, for example between a windowpane and a person's skin. Heat always radiates from the warmer to the colder surface. Hence, the human body can either lose or gain heat through radiation.

The amount of radiated heat depends primarily on the temperature difference between the two surfaces but not on the temperature of the air between them.

The amount of radiating energy Q_R lost or gained by the body through radiation depends essentially on the size S of the participating body surface and on the difference Δ between the quadrupled temperatures T (in degrees Kelvin) of the participating surfaces:

$$Q_R = f(S, \Delta T_4) \qquad (9.3)$$

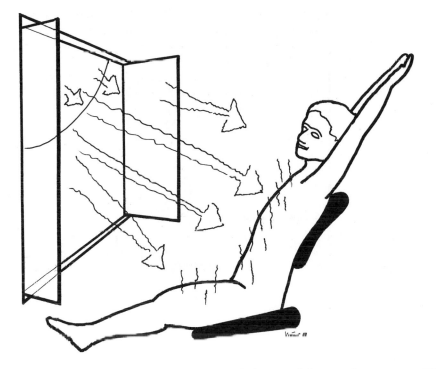

Figure 9.1 Heat radiated from the sun can feel so good, because it warms us. Of course, if we are already hot, we seek the shade.

This can be approximated by

$$Q_R \approx S \times h_R \Delta t \tag{9.4}$$

where h_R is the heat-transfer coefficient and t the temperature in °C. (For more detail, read Kroemer et al. (1997) or Youle (1990).)

Figure 9.1 illustrates how heat from the sun warms our skin, without directly affecting the temperature of the air. Heat may also radiate from a fire, a hot window pane, a heating radiator, or any other warm body whose surface temperature is higher than that of our skin. Of course, if the skin is warmer than, say, the cold glass pane of a window, and if there is no insulator such as clothing between the skin and the glass, then we radiate our heat to the window, as sketched in Figure 9.2. In this case, our body is losing heat, and we are growing colder.

Heat exchange through convection, C, and conduction, K, both follow the same thermodynamic rules. The heat transferred is again proportional to the area of human skin participating in the process, and to the temperature difference between skin and the adjacent layer of the external medium. Hence, in general terms, heat exchange by convection or conduction is

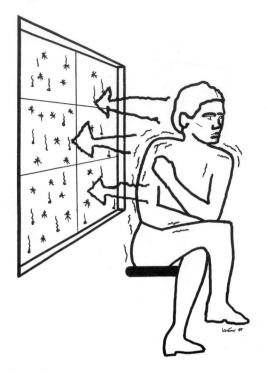

Figure 9.2   Our skin radiates heat to a colder surface, such as a cold windowpane in the winter. To avoid losing heat, we can cover up – either our body with clothing, or the glass with a curtain; or both.

$$Q_{C,K} = f(S, \Delta t) \tag{9.5}$$

approximated by

$$Q_{C,K} \approx S \times k \times \Delta t \tag{9.6}$$

with k the coefficient of conduction or convection.

Heat exchange through convection, C, takes place when the human skin is in contact with air (or other gas or fluid). Heat energy transfers from the skin to the adjacent layer of colder gas (or fluid), or transfers to the skin if the surrounding medium is warmer.

Convective heat exchange is facilitated if air moves quickly along the skin surface, which helps maintain a temperature (and humidity) differential. This movement can be enhanced by an air fan.

Figure 9.3 shows how the air moving along our skin carries away the layer of warmed (and moist) air that is close to our skin. This makes convective heat exchange more effective and, if the air around us (or the water, when we swim) is colder than the skin, we lose body heat. Keeping the air layer intact, such as when wearing loose-fitting clothes, acts as an

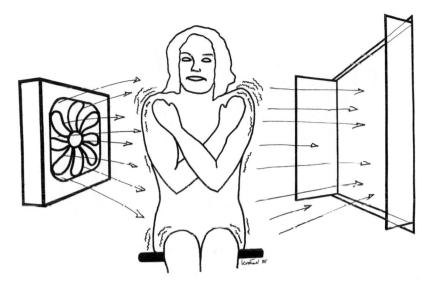

Figure 9.3   Air flow removes the layer of warmed air from our skin and makes us lose heat by convection to the colder fresh air. Stopping the current of air, and using clothes that insulate (partly by keeping a warm layer of air trapped) prevents us from growing cold via convection.

insulator reducing the heat transfer. The same thermodymanic mechanisms are in effect when we are hot and sweaty and enjoy the cooling effect of air being blown at us through a fan or a refreshing breeze, as sketched in Figure 9.4.

Conductive heat exchange, K, exists when the skin is in touch with a solid body. As long as there is a difference in temperature, heat flows toward the colder body, strongly so if the conductance of the piece is high; less energy flows if the skin touches an insulator having a low k value.

The wood of furniture feels warm because its heat-conduction coefficient is below that of human tissue. Metal of the same temperature accepts body heat easily and conducts it away; therefore, it feels colder than wood.

Heat exchange by evaporation, F, is in only one direction: the human loses heat by evaporation. There is never condensation of water on living skin, which would add heat. Evaporation of sweat, which is mostly water, requires energy of about 580 cal/cm^3; this energy is mostly taken from the body and hence reduces the heat content of the body by that amount. Some water is evaporated in the respiratory passages but most appears (as sweat) on the skin.

There is always some perspiration and hence sweat evaporation going

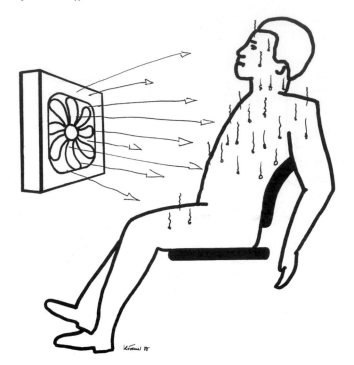

Figure 9.4   Air flow removes the layer of moist and warm air that our hot body created from our skin and makes us lose heat by evaporation of sweat, and by convection, to the drier and colder air blown in. This breeze is welcome on a hot day; not so welcome on a brisk one.

on; that is why our clothes smell when worn too long, even if we are not engaged in strenuous activity while wearing the clothing.

The heat lost by evaporation Q_E from the human body depends mostly on the participating wetted body surface, S, and on air humidity, h:

$$Q_E \approx f(S, h) \tag{9.7}$$

Higher humidity makes evaporative heat loss more difficult than in dryer air conditions. Movement of the air layer at the skin increases the actual heat loss through evaporation since it replaces humid air by drier air. The warmer the air, the more water vapor it can accept.

Now we can understand why using a fan in a hot climate, which blows air at us, feels so refreshing (Figure 9.4). The air flow takes away the layer of humid air around our body, which we have pretty much saturated with our evaporated sweat, and replaces it with dryer air. While the blown air is still hot, and while overtly we would prefer a cool breeze, warm air accepts our sweat more easily.

Because air's saturation with water depends on air temperature (besides atmospheric pressure), heat loss by evaporation is indirectly also a function of Δt, as are the other heat transfers by radiation, convection and conduction.

Heat balance

The actual amounts of heat exchanged with the surroundings depend, directly or indirectly, on the difference in temperature between participating body surfaces and the environment, as the equations listed above show. The body keeps its core temperature constant when it achieves heat balance; this occurs when the metabolic energy H developed in the body, see Equation (9.2), is in equilibrium with the heat exchanged with the environment by radiation R, convection C, conduction K, and evaporation E. Equilibrium is achieved if they all add up to zero.

This condition of heat balance can be expressed as

$$H \pm R \pm C \pm K - E = 0 \qquad (9.8)$$

When the body loses energy to the environment, one, several or all of the quantities R, C, and K are negative, but positive if the body gains energy from the surroundings. E is always negative.

Temperature regulation and sensation

As mentioned earlier, if our body's core temperature changes about 2° from its 37°C norm, our body's functions and its ability to perform tasks are severely and negatively affected, larger temperature deviations can kill us. Even in a sheltered office environment, where the human thermoregulatory system can easily keep the core close to 37°C, there are major temperature differences from one body region to another.

In the human body, heat is generated in metabolically active tissues – primarily by skeletal muscles, but also in internal organs, fat, bone, and connective and nerve tissue. The heat energy is circulated throughout the body by the blood. Blood vessel constriction, dilation, and shunting regulate the blood flow within the body. Heat exchange with the environment takes place at the body's respiratory surfaces (primarily in the lungs) but mostly through the skin. The equations shown above indicate that the energy exchange with our surroundings depends on the area of participating skin surface, especially skin that is naked and exposed to the environment. Clothing helps insulate the skin surface from heat transfer.

In a cold environment, body heat must be conserved. The body does this automatically by reducing blood flow to the skin (which keeps the skin colder) and we do it deliberately by increasing insulation via clothing. In a hot environment, body heat must be dissipated and heat gain from the environment must be minimized. The body does this by increasing the blood flow to the skin, making it warmer, and by increased sweat production and evaporation – and we help this process deliberately by dressing in lighter clothes.

> The healthy body naturally responds to a cold environment by making its skin surface colder, and to a hot environment by making the skin warmer.

If the body is about to overheat, internal heat generation must be diminished. For this to happen, muscular activities must be reduced. Consider the warm-climate countries in which inhabitants routinely take mid-day breaks (a siesta). Such breaks evolved in large degree because people recognized their bodies' signals and rested muscles accordingly. Their bodies are overheated due to the peak heat of the day and the likely ingestion of a mid-day meal; taking a siesta allows the body to sharply reduce muscular activity (through sleep). In the opposite case, when more heat must be generated, increasing our work or exercise level and thus augmenting our muscular activities will help us warm up.

> If you have ever watched a foot race like a 10-K or a marathon held in cold weather, you will notice that the joggers dress in layers so that they can discard clothing as they warm up during their run. In fact, in highly competitive or elite races, many runners dress quite lightly and likely feel very chilly at the commencement of their run because they recognize that heat generation will ensue within a mile or two. Another illustration is the common sight of a downhill skier hopping or stomping his feet while waiting in a queue for the ski lift on a cold winter's day.

Muscular, vascular, and sweat-production functions cooperatively regulate the body heat content in response to the surrounding climate.

Various temperature sensors are located in the core and the shell of the body. Heat sensors generate signals particularly in the range of approximately 38–43°C. Cold sensors are most sensitive from about 35 to 15°C. There is some overlap in the sensations of cool and warm in the intermediate range. Between about 15 and 45°C, perception of either cold or hot condition is highly adaptable; under suitable circumstances, we can get used to these temperatures. Below 15 and above 45°C the human temperature sensors are less discriminating but also less adapting. A paradoxical effect occurs around 45°C, where sensors again signal cold (we may get goose bumps) while in fact the temperature is rather hot.

Achieving thermal homeostasis

In a sheltered environment, such as an office, the human body achieves thermal equilibrium (called homeostasis) primarily by regulating the blood flow from deep tissues and muscles to skin and lungs. Furthermore, muscle activities generate heat: *involuntarily* by shivering, *voluntarily* by exercise or work efforts. If the goal is to achieve heat gain, we initiate muscle contractions; but if heat gain must be avoided, we reduce or abolish muscular activities.

Clothing at work

Changes in clothing and in shelter climate represent our conscientious actions to establish thermal homeostasis. They achieve (together with blood-flow regulation and muscle activities) the appropriate temperature gradient between the skin and the environment. They affect radiation, convection, conduction, and evaporation. Light or heavy clothing has different permeability and ability to establish stationary insulating layers. Clothes affect conductance, i.e., energy transmitted per surface unit, time, and temperature gradient. Also, their color determines how much external radiation energy is absorbed or reflected. Especially in the sun, light-colored clothes reflect radiated heat while dark surfaces absorb it.

The insulating value of clothing is measured in clo units, with 1 clo $=$ $0.155°C\ m^2\ W^{-1}$. This is approximately the value of the normal clothes worn by a sitting person at rest in an office where the air is at about 21°C and has about 50 per cent relative humidity.

Assessing the thermal environment

In terms of physics and engineering, our thermal office environment is described by four factors:

* air temperature
* humidity
* air movement and
* temperature of surfaces that exchange energy by radiation and conduction

The combination of these four factors determines the physical conditions of the climate and our perception of the climate.

Air temperature is measured with thermometers, thermistors, or thermocouples. For exact measurements we must ensure that the ambient temperature is not affected by the other three climate factors, particularly by humidity. To determine the so-called dry temperature of ambient air, one keeps the sensor dry and shields it with a surrounding bulb that reflects radiated energy. Hence, air temperature is often measured with a so-called dry-bulb thermometer.

Air humidity may be measured with a psychrometer, hygrometer, or other electronic devices. These usually rely on the fact that the cooling effect of evaporation is proportional to the humidity of the air; higher vapor pressure makes evaporative cooling less efficient. Therefore, we can measure humidity using two thermometers, one dry, one wet.

The highest absolute content of water vapor in the air is reached just before water droplets develop. The amount of possible vapor depends on pressure and temperature of the air; lower pressure and higher temperature allow more water vapor than lower temperatures and higher pressure. When one speaks of humidity in per cent, one refers to relative humidity, the actual percentage of vapor content in relation to the possible maximal content (absolute humidity) at the given air temperature and air pressure.

Air movement is measured with various types of anemometers using mechanical or electrical principles. We can measure air movement also with two thermometers – one dry and one wet (similar to what can be done to assess humidity), relying on the fact that the wet thermometer shows more increased evaporative cooling with higher air movement than the dry thermometer. Air moving at higher velocity cools human skin better (by convection and evaporation) than stagnant air.

Radiant heat exchange depends primarily on the difference in temperatures between the participating surfaces of the person and the surroundings, on the emission properties of the radiating surface, and on the absorption characteristics of the receiving surface. One practical way to assess the amount of energy acquired through radiation is to place a thermometer inside a black globe that absorbs practically all arriving radiated energy.

Thermal comfort

A person's feeling of comfort is not only determined by the physics of thermal balance, discussed above. Instead, other factors that influence our perception of physical comfort and wellbeing include skin dampness in a warm environment and skin temperature in a cold environment.

In the past, various techniques were used to assess the combined effects of some or all of the four environmental factors and to express these in one model, chart, or index. They resulted in several empirical thermal indices, which are based on data compiled from the statements of subjects who were exposed to various climates. Most establish a reference or effective climate that feels the same as various combinations of the several climate components.

A well-known example is the effective temperature (ET); it reflects combinations of dry temperature, humidity, and air movement and applies to the office environment where people wear rather light clothing and work at fairly low levels of physical effort. Such a climate index can be provided by instruments specially arranged to respond to climate components the way a human being would.

The wet-bulb globe temperature (WBGT) index is generated by an instrument with three sensors whose readings are automatically weighted and then combined. It weights the combined effects of climate parameters as follows:

Outdoors it is

$$WBGT = 0.7WB + 0.2GT + 0.1DB \qquad (9.9)$$

Indoors it is

$$WBGT = 0.7WB + 0.3GT \qquad (9.10)$$

where WB is the wet-bulb temperature of a sensor in a wet wick exposed to natural air current, GT is the globe temperature at the center of a black sphere of 15 cm diameter, and DB is the dry-bulb temperature measured while shielded from radiation.

The Bedford and the ASHRAE scales are widely used to assess thermal comfort; they yield similar results (Youle, 1990).

Jim, a recent college graduate from Southern Florida, has just relocated to Chicago. On a blustery day in February, just a few days after moving to the Midwest, Jim has an interview with a CPA firm downtown. As he dresses in his best business suit that morning, he turns on the Weather Channel to get the forecast. He faces a 30-minute commute on public transportation, and he wants to make sure he selects the most appropriate clothing. After all, since he is most accustomed to the sunny and warm climes of Southern Florida, he wants to avoid freezing as he makes his way downtown. As Jim finishes dressing, he hears the cheerful weather forecaster predict 35–40°F, not bad for Chicago in the dead of winter. He throws on an overcoat but forgoes hat, scarf, and gloves, thinking that the coat will suffice. By the time he walks to the elevated train stop and stands on the platform for 10 minutes, awaiting the train, he is shivering and shaking from the cold. The whistling wind is howling around him as he blows on his freezing fingers and huddles closer to the heating lamps on the platform. The woman next to him watches his gyrations, amused, and takes in his lightweight overcoat. Jim notices her perusal. "I made a point of listening to the forecast this morning," he grumbles, "and they said it would be 35°F. This feels a lot colder than that." She nods. "Well, sure, it is a lot colder than that. It's got to be no more than 10° when you factor in the windchill." She sees the blank look on Jim's face. "Come on, now. Haven't you ever heard of windchill factors?" the woman asks with a chuckle. "As a matter of fact," Jim responds dryly, "Being from Miami, I have not."

Reactions to hot environments

In hot environments, the body produces heat and must dissipate it. To achieve this, the skin temperature should be above the immediate environment in order to facilitate energy loss through convection, conduction, and radiation.

If heat transfer is not sufficient, sweat glands are activated, and the evaporation of the produced sweat cools the skin. Recruitment of sweat glands from different areas of the body varies among individuals. The overall amount of sweat developed and evaporated depends very much on clothing, environment, work requirements, and also the individual's acclimatization (discussed below). Also, some of us just sweat more than others. If much sweat is evaporated, drinking fluid, best just cool water, replaces the fluid our body has lost.

If heat transfer by sweat evaporation is still insufficient, muscular activities must be reduced to lower the amount of energy generated. This is the final and necessary action of the body when it is approaching a point where the core temperature would exceed a tolerable limit. As we mentioned earlier, taking a midday siesta during the hottest period of the day in a hot climate is a good example of a natural escape from overheating, especially when resting in the cool shade.

If the body has to choose between unacceptable overheating and continuing to perform physical work, the choice will be in favor of core-temperature maintenance, which means reduction or cessation of work activities.

Smells and odorants

Especially in a hot environment, smells can be a problem. Most women are better able than men to perceive odors. What we call odor is the result of sensations generated by odorants, which are chemical substances. (Some substances are odorless, such as carbon monoxide.) Most everyday odorants are mixtures of several or many basic chemical components, which generate complex odors to which different people, in different environments, and at different lengths of exposure react quite differently. Many odorants (often accompanying air contaminants) stem from the outgassing of building materials, from byproducts of running machines, from the industrial and traffic environment, and from human activity like smoking or consuming pungent food. Not all odorants are external: the body also generates smells. Body odor arises from the apocrine glands, which are small in infants but develop during puberty. Most of these glands are located at the armpits, face, chest, genitals, and anus.

While we use our olfactory sense daily, even today little is known systematically about it, partly because a smell is not easily quantified.

Commonly, its sensation is described in relation to other smells, pleasant or not. Assessment of odor annoyance has been attempted in a manner similar to noise classification (Hangartner, 1987). One the first quantitative studies was done in Danish offices; very little smell pollution came from people (who, in case you are wondering, bathed on average 0.7 times a day and did not smoke) but far more emanated from smokers, materials in the office, and the ventilation system (Fanger, 1988a).

The effects of odors can be physiological, independent of the actual perception. When strong, they may stimulate the central nervous system, eliciting changes in body temperature, appetite and arousal. Such odors may trigger responses including nausea, headache, coughing; irritation of eyes, nose, and throat. Odors also can have psychological and psychogenic effects that concern especially attitude and mood, including anger or benevolence toward others, cooperation, creativity, self-perception, and performance. They may be part of the sick building syndrome (Ballard, 1995; Fanger, 1988b).

Increased humidity reduces odor intensity; to this end, vapor pressure in the office air should be between about 10 and 15 Torr, and about 10 liters of outside (fresh) air should be provided per second for each person in the office (ASHRAE, 1997, 1999a). Building codes and standards (such as ASHRAE, 1999b) require minimal ventilation rates to dilute contaminants and odorants, but they do not consider air temperature and humidity as important for perceived air quality which, as is well known, declines as either or both increase.

Reactions to cold environments

The human body has few natural defenses against a cold environment. Most of our counter-actions are behavioral, such as putting on suitably heavy clothing or using external sources of warmth.

In a cold climate, the body must conserve heat. To this end, the temperature of the skin is naturally lowered, reducing the temperature difference against the outside. Keeping the circulating blood closer to the core, away from the skin, accomplishes this; for example, the blood flow in the fingers may be reduced to 1 per cent of that in a moderate climate. This results in cold fingers and toes.

The development of goose bumps on the skin helps to retain a relatively warm layer of stationary air close to the skin. The stationary envelope acts like an insulator, reducing energy loss at the skin.

The other major reaction of the body to a cold environment is to increase metabolic heat generation. This can be done by purposeful muscular activities, such as in moving body segments like flexing the fingers. Such dynamic muscular work may easily increase the metabolic rate to ten or more times the resting rate.

Metabolic heat generation often occurs involuntarily through shivering. The onset of shivering is normally preceded by an increase in overall muscle tone in response to body cooling. Right before we shiver, we usually experience a general feeling of stiffness. Then suddenly shivering begins, caused by muscle units firing at different frequencies of repetition and out of phase with each other. Since no mechanical work is done on the outside, the total activity is transformed into heat production, allowing an increase in the metabolic rate to up to four times the resting rate. Shivering usually begins in the neck, apparently because warmth is important to supply blood to the brain.

Do drafts make us sick? Some people believe they do; for example, when in Germany somebody shouts "es zieht!" (a draft!) then the offending window is shut, even on a hot day; no questions asked. Perhaps your grandmother even advised you to steer clear of drafts in order to avoid getting a cold. But do you really suffer terrible health consequences from a draft, as these people believe? Not likely. Moving air can cool us, and we may need the cooling. Movement of clean air does not make you sick, and even getting wet and cold does not make you catch a cold.

However, contaminated air can cause health problems; somebody with a cold, or the flu, breathing and sneezing on you can indeed infect you. In the office, however, the likelihood of airborne person-to-person respiratory infection is reduced by increased air flow, and the air can be disinfected by filtering or irradiation.

In short, Grandma was wrong (in this case only, of course); drafts do not make you sick. Contamination and viruses do. (Adapted from Kroemer et al., 2001.)

Acclimatization

Continuous or repeated exposure to hot and (to a lesser degree) cold conditions brings about a gradual adjustment of body functions, resulting in a better tolerance of the climate stress.

Acclimatization to heat is characterized by increased sweat production, lowered skin and core temperature, and a reduced heart rate, compared with a person's first reactions to heat exposure. The process (called acclimation) is very pronounced within about a week, and full acclimatization is achieved within about 2 weeks. Interruption of heat exposure of just a few days reduces the effects of acclimation; upon return to a moderate climate, acclimatization is entirely lost after about 2 weeks.

A healthy person can adjust to some heat, dry or humid. Heat acclimatization does not depend on the type of work performed, whether heavy and short or moderate but continuous. A person who is healthy and well trained acclimates more easily than a person who is in poor physical condition, but training cannot replace acclimatization. Since the body can adapt to heat, but not to dehydration, liberal drinking of water is helpful, both during acclimation and then throughout heat exposure, to replace fluid lost by evaporation of sweat.

Joann works as an administrative assistant at a brewery in Milwaukee. It has been a long, cold winter, and now, in March, it still seems like the chilly, snowy weather in Wisconsin will last for at least several more weeks. For once, though, Joann does not mind the lingering winter; she's just left freezing Milwaukee and arrived in Mexico for a 7-day honeymoon with her new husband, Jake. After a leisurely brunch at their luxurious hotel on their first full day, they decide to take a long walk into town and back. Within 20 minutes of their stroll in the midday sun, however, Joann is feeling quite winded and very warm. She calls out to her husband, who is walking in front of her, to slow down. Just then, she notices that they are passing by a crew of laborers working to lay the foundation of a new hotel close to town. She notes with some amazement that the men on the crew are working seemingly tirelessly, shoveling dirt and effortlessly carrying and moving large pieces of building material. She points out the crew to her husband, and the two pause and watch the hard-working laborers for several minutes. Joann catches her breath as they watch and wonders out loud how the men are capable of functioning like that in what seems to her to be oppressive heat. "I guess," Jake surmises, "they're just used to it."

Acclimatization to cold is much less pronounced; in fact, there is some doubt that true physiological adjustment to moderate cold actually takes place when appropriate clothing is worn.

As in a hot climate, the body must maintain its core temperature near 37°C in a cold environment. When exposed to cold, the human body first responds by peripheral vasoconstriction (constriction of blood vessels), which lowers skin temperature, in order to decrease heat loss through the skin. This can lead to local acclimatization in terms of the flow of blood in the hands and face. However, the blood supply to the brain remains intact, so up to 25 per cent of the total heat loss can take place at the warm surfaces of the neck and head. The predominant adjustment to cold conditions is in choosing proper clothing and work with the result that, in normally cold temperatures, the body has little or no need to change its rate of heat production or, relatedly, of food intake. In other words, acclimatization to cold temperatures appears to be the result primarily of people becoming smarter at dealing with it – by learning to dress appropriately, for example.

There are no great differences between females and males with respect to their ability to adapt to either hot or cold climates, with women possibly at a slightly higher risk for heat exhaustion and collapse and for cold injuries to extremities. However, these slight statistical tendencies can be easily counteracted by ergonomic means and may not be obvious at all when observing only a few persons of either gender.

Effects of heat or cold on mental performance in the office

It is difficult to evaluate the effects of heat (or cold) on mental or intellectual performance. This is because of the lack of practical yet objective testing methods coupled with the huge range of subjective variations among individuals in an office in which the climate fluctuations are rather small. However, as a rule, mental performance deteriorates with rising room temperatures, starting at about 25°C for the non-acclimatized person. That threshold increases to 30 or even 35° if the individual has acclimatized to heat. Brain functions are particularly vulnerable to heat; keeping the head cool improves the tolerance to elevated deep body temperature. A high level of motivation may also counteract some of the detrimental effects of heat. Thus, in laboratory tests, mental performance is usually not significantly affected by heat as high as 40°C WBGT (Kroemer et al., 1997, 2001).

Extreme cold conditions are not often found in offices. In the unlikely case of body core temperature dropping below the 35°C, vigilance and mental performance are reduced and nervous coordination suffers. Manual dexterity is reduced if finger skin temperatures fall below 20°C. Tactile sensitivity is reduced at about 8°C, and near 5°C, skin receptors for pressure and touch cease to function; the skin feels numb at these low temperatures.

Air-conditioning of offices

Throughout the 1800s, the architecture of large office buildings, and the arrangement of the offices within, still followed the example of the Uffici (Offices) in Florence, Italy, constructed between 1560 and 1581. They were built in U-form around an open court so that every room had a window to the outside, to provide natural lighting and ventilation.

One of the authors (Karl K.) experienced, with a bit of longing and nostalgia, the immigration office in Bombay (now Mumbay), India, in 1983: a long, two-story building with wide verandahs in a park-like setting, all windows were open to the breezes – which made paperweights necessary to keep the stacks of documents in place.

In such old large buildings, heating was done, if needed, by coal fires and stoves in the rooms, and bad odors abounded when the windows had to be kept closed. Although electricity was widely available about 1890, and some mechanical ventilation of the lower floors of tall office buildings was already in use by that time, design plans in forms of U, O, E, I, H, and T were employed into the 1930s to allow windows to the outside in every room (Arnold, 1999).

In 1906, Frank Lloyd Wright's Larkin Administration Building was constructed. It had sealed ventilation with some cooling and heating of the air. This was an early form of air-conditioning, a term that came to be generally used in the 1920s while the practice became commonplace in office buildings in the USA only during the following decade. In 1936, the Milam building in San Antonio, TX, was the first to be fully air-conditioned. This technology allowed, city-block wide edifices (Arnold, 1999).

In 1932, the PSFS building in Philadelphia was erected in the new International Style created by Le Corbusier in France. It was fully air-conditioned, which the contemporary Empire State and RCA buildings in New York were not. Other fully conditioned buildings were the Hershey Chocolate headquarters in Pennsylvania and the Detroit Edison building. The one edifice that really utilized all the new technology was Frank L. Wright's Johnson's Wax Administration Building, completed in 1939. It was entirely sealed and fully air-conditioned, with under-floor heating (Arnold, 1999).

Air-conditioning was increasingly used in large office buildings, which, coupled with electric lighting, led to large and windowless buildings. The architectural and engineering evolution was interrupted by the Second World War, after which it resumed at full speed. Air-conditioning of large and small office buildings, even of single rooms by (often noisy) window air-conditioners, became commonplace in North America. In the course of this development, many concerns grew as well: about the effects of missing natural light on health and performance (Hedge, 2000); about the spread of tobacco smoke and other air contaminants, even diseases, by forced air flow (Fanger, 1988a,b); about acceptable indoor air quality (ASHRAE Handbooks; Kuehn et al., 1998; Schiller and Arens, 1988); about the work performance in offices that are too cool or too warm; and about the effects of putting many persons together in large rooms in terms of supervision, performance, and behavior (see Chapters 2 and 3).

ERGONOMIC DESIGN RECOMMENDATIONS

There are many ways to generate a thermal environment that suits the physiological needs of people as well as their individual preferences. The primary approach is to adjust the physical conditions of the climate (temperatures, humidity, and air movement) which in turn influence the heating or cooling of the body via radiation, convection, conduction, and evaporation. The body's interactions are complex and must be carefully considered when designing and controlling the environment; guidance may be taken from ASHRAE recommendations in the USA, from international standards such as those outlined by ISO, and from national regulations and regional recommendations and customs.

> What is of importance to the individual is not the climate in general, the so-called macroclimate, but the climatic conditions with which one interacts directly. Every person desires an individual microclimate that feels comfortable under given conditions of adaptation, clothing, work and individual preference.

The suitable microclimate is not only highly individual but also variable. It depends, for example, on age; older persons tend to be less active and to have weaker muscles, to have a reduced caloric intake, and to start sweating at higher skin temperatures. It depends on the surface-to-volume ratio, which in children is much larger than in adults, and on the fat-to-lean body-mass ratio, generally larger in women than in men.

Thermal comfort depends also on the type and intensity of work performed. Physical work in the cold leads to increased internal heat production and hence to decreased sensitivity to the cold environment, while intensive physical work in a hot climate can become intolerable if an energy balance cannot be achieved, conditions not often found in offices.

Of course, clothing also affects the microclimate. Clothing worn determines the surface area of exposed skin. More exposed surface areas allow better dissipation of heat in a hot environment but can lead to excessive cooling in the cold. Air bubbles contained in the clothing material or between clothing layers provide insulation, both against hot and cold environments. This is why wearing appropriate layered clothing can help us stay warm in the winter. Permeability to fluid (sweat) and air plays a role in heat and cold. Even the colors of the clothing we choose play a role in regulating our temperature; in a heat-radiating environment, such as in sunshine, we will feel more comfortable in lighter-colored clothes. Darker colors absorb radiated heat while light colors reflect incident energy.

Convection heat loss is increased if air moves swiftly along exposed surfaces. Therefore, with increased air velocity, body cooling becomes more pronounced. On a warm day, you will likely feel more comfortable if a breeze exists rather than if the air is stagnant, even though the temperature itself is unchanged. Similarly, on a very cold day, wind chills can make us feel significantly more miserable than if the air were still.

Old non-air-conditioned offices had to rely in the summer on airflow through open windows and doors to keep the people cool enough. In those offices, people knew by experience how to dress and behave appropriately to deal with their climate – they welcomed drafts but kept papers weighted down so they didn't get blown away. In the modern office, the stream of conditioned air can be quite uncomfortable if, due to poor design or placement of the air ducts, it blasts directly on you. The flow of cold air can chill you thoroughly – although not as badly as indicated by the "wind chill temperatures" that weather forecasters love to quote to make their announcements more dramatic. Note that these wind chill temperatures are based on the cooling of naked human skin, and that most of the air temperatures and velocities shown do not occur in offices.

Thermal comfort is also affected by acclimatization, a state in which the body (and mind) have adjusted to changed environmental conditions. A climate that felt rather uncomfortable during the first day of exposure may be quite agreeable after a couple of weeks. Seasonal changes in climate, unusual work, different clothing, and an evolving attitude all have major effects on what we are willing to accept or even to consider comfortable. In the summer, most people find warm, windy, and rather humid conditions comfortable while during the winter we feel that cool and dry weather is normal. This is what many of us are accustomed to.

For these reasons, various and variable combinations of the climate factors of temperature, humidity, and air movement can subjectively appear quite similar. The WBGT discussed earlier is most often used to assess the effects of warm or hot climates on the human; for a cold climate, various similar approaches have been proposed but are not universally accepted yet; see Youle (1990) for a critical overview and for details.

Ergonomic recommendations

This feels good

With appropriate clothing and light physical work in the office, comfortable environment temperature ranges from about 21°C to 27°C in a warm climate or during the summer, but lower – between 18°C and 24°C – in a cool climate or during the winter.

The difference between air temperatures at floor and head levels should be less than about 6°C. Differences in temperatures between body surfaces and surfaces on the side (such as walls or windows) should not exceed approximately 10°C.

The preferred range of relative humidity is from 30 to 70 per cent, best near 40–50 per cent.

Air velocity should not exceed 0.5 m/s, preferably remaining below 0.1 m/s. Air flow should not generate sound levels above 60 dB.

Deviations from these zones are uncomfortable, can make work difficult, and may even become intolerable.

If the sun shines onto employees, particularly on warm days, they should be able to move out of the sun, or to get into the shadow of blinds, curtains or screens.

What to do if you are not comfortable

If you feel too warm:

- lower the room temperature
- move away from a heat source such as a radiator, a warm wall or window; get out of the sun
- move closer to a cool surface
- lower air humidity (use a de-humidifier)
- increase the air movement around you (unless it is very hot air)
- wet your exposed skin; place a cool/moist piece of cloth on your forehead, neck or wrist
- take off a layer of clothing; bare more skin
- keep your body at rest, do not exercise it

If you feel too cool:

- increase the room temperature
- move closer to a heat source such as a radiator, a warm wall or window; move into the sunshine
- move away from a cool surface
- move closer to a warm surface
- decrease the air movement around you (unless it is nicely warm air)
- add a layer of clothing; cover more skin
- keep your body moving

If you feel too dry (dry throat, nose):

- increase air humidity by evaporating water on a warm surface, using a humidifier; also drink water

If you draw sparks of static electricity:

- increase air humidity by evaporating water on a warm surface, using a humidifier; also check your clothing including your shoes because they may generate an electric charge against the office furniture or carpeting

References

Arnold, D. (1999). The Evolution of Modern Office Buildings and Air Conditioning. *ASHRAE Journal 41*(6), 40–54.

ASHRAE (1997). *ASHRAE Handbook: Fundamentals.* Atlanta, GA: American Society of Heating, Refrigerating and Air-Conditioning Engineers.

ASHRAE (1999a). *ASHRAE Handbook: HVAC Applications.* Atlanta, GA: American Society of Heating, Refrigerating and Air-Conditioning Engineers.

ASHRAE (1999b). *ASHRAE Standard 62-1999, Ventilation for Acceptable Indoor Air Quality.* Atlanta, GA: American Society of Heating, Refrigerating and Air-Conditioning Engineers.

Ballard, B. (1995). How Odor Affects Performance: A Review, in *Proceedings, ErgoCon '95, Silicon Valley Ergonomics Conference and Exposition.* San Jose, CA: San Jose State University, 191–200.

DiNardi, S. R. (Ed.) (1997). *The Occupational Environment Its Evaluation and Control.* Fairfax, VA: American Industrial Hygiene Association.

Fanger, P. O. (1988a). The Olf and the Decipol. *ASHRAE Journal 30*(10), 35–38.

Fanger, P. O. (1988b). Hidden Olfs in Sick Buildings. *ASHRAE Journal 30*(11), 40–43.

Hangartner, M. (1987). Standardization in Olfactometry with Respect to Odor Pollution Control; Assessment of Odor Annoyance in the Community. *Presentations 87-75A.1 and 87-75B.3 at the 80th Annual Meeting of the APCA.* New York.

Hedge, A. (2000). Where Are We in Understanding the Effects of Where We Are? *Ergonomics 43*, 1019–1029.

Karwowski, W. and Marras, W. S. (Eds.) (1999). *The Occupational Ergonomics Handbook.* Boca Raton, FL: CRC Press.

Kroemer, K. H. E., Kroemer, H. J. and Kroemer-Elbert, K. E. (1997). *Engineering Physiology: Bases of Human Factors/Ergonomics* (3rd ed.). New York: Van Nostrand Reinhold–Wiley.

Kroemer, K. H. E., Kroemer, H. B. and Kroemer-Elbert, K. E. (2001). *Ergonomics: How to Design for Ease and Efficiency* (2nd ed.). Upper Saddle River, NJ: Prentice Hall.

Kuehn, T. H., Ramsey, J. W. and Threlkel, J. L. (1998). *Thermal Environmental Engineering* (3rd ed.). Englewood Cliffs, NJ: Prentice Hall.

Plog, B. A. (Ed.) (2001). *Fundamentals of Industrial Hygiene* (5th ed.). Itasca, IL: National Safety Council.

Salvendy, G. (Ed.) (1997). *Handbook of Human Factors and Ergonomics* (2nd ed.). New York: Wiley.

Schiller, G. E. and Arens, E. A. (1988). Thermal Comfort in Office Buildings. *ASHRAE Journal 30*(10), 26–32.

Youle, A. (Ed.). (1990). *The Thermal Environment (Technical Guide No. 8, British Occupational Hygiene Association).* Leeds: Science Reviews Ltd. and H and H Scientific Consultants Ltd.

Index